Design Education for Fostering Creativity and Innovation in China

Kin Wai Michael Siu
The Hong Kong Polytechnic University, Hong Kong

Giovanni J. Contreras
The Hong Kong Polytechnic University, Hong Kong

A volume in the Advances in Higher Education and Professional Development (AHEPD) Book Series

IGI Global
DISSEMINATOR OF KNOWLEDGE
www.igi-global.com

Published in the United States of America by
IGI Global
Information Science Reference (an imprint of IGI Global)
701 E. Chocolate Avenue
Hershey PA 17033
Tel: 717-533-8845
Fax: 717-533-8661
E-mail: cust@igi-global.com
Web site: http://www.igi-global.com

Library of Congress Cataloging-in-Publication Data

Names: Siu, Kin Wai Michael, editor. | Contreras, Giovanni J., editor.
Title: Design education for fostering creativity and innovation in China / Kin Wai Michael Siu and Giovanni J. Contreras, editors.
Description: Hershey : Information Science Reference, 2016. | Includes bibliographical references and index.
Identifiers: LCCN 2016029354| ISBN 9781522509110 (hardcover) | ISBN 9781522509127 (ebook)
Subjects: LCSH: Technological innovations--China. | Creative ability in business--China. | Industrial design--Study and teaching--China.
Classification: LCC HC430.T4 D47 2016 | DDC 338/.0640951--dc23 LC record available at https://lccn.loc.gov/2016029354

This book is published in the IGI Global book series Advances in Higher Education and Professional Development (AHEPD) (ISSN: 2327-6983; eISSN: 2327-6991)

British Cataloguing in Publication Data
A Cataloguing in Publication record for this book is available from the British Library.

Advances in Higher Education and Professional Development (AHEPD) Book Series

ISSN: 2327-6983
EISSN: 2327-6991

Mission

As world economies continue to shift and change in response to global financial situations, job markets have begun to demand a more highly-skilled workforce. In many industries a college degree is the minimum requirement and further educational development is expected to advance. With these current trends in mind, the **Advances in Higher Education & Professional Development (AHEPD) Book Series** provides an outlet for researchers and academics to publish their research in these areas and to distribute these works to practitioners and other researchers.

AHEPD encompasses all research dealing with higher education pedagogy, development, and curriculum design, as well as all areas of professional development, regardless of focus.

Coverage

- Adult Education
- Assessment in Higher Education
- Career Training
- Coaching and Mentoring
- Continuing Professional Development
- Governance in Higher Education
- Higher Education Policy
- Pedagogy of Teaching Higher Education
- Vocational Education

IGI Global is currently accepting manuscripts for publication within this series. To submit a proposal for a volume in this series, please contact our Acquisition Editors at Acquisitions@igi-global.com or visit: http://www.igi-global.com/publish/.

The Advances in Higher Education and Professional Development (AHEPD) Book Series (ISSN 2327-6983) is published by IGI Global, 701 E. Chocolate Avenue, Hershey, PA 17033-1240, USA, www.igi-global.com. This series is composed of titles available for purchase individually; each title is edited to be contextually exclusive from any other title within the series. For pricing and ordering information please visit http://www.igi-global.com/book-series/advances-higher-education-professional-development/73681. Postmaster: Send all address changes to above address.

Titles in this Series

For a list of additional titles in this series, please visit: www.igi-global.com

Integrating Video into Pre-Service and In-Service Teacher Training
Pier Giuseppe Rossi (University of Macerata, Italy) and Laura Fedeli (University of Macerata, Italy)
Information Science Reference • copyright 2017 • 340pp • H/C (ISBN: 9781522507116) • US $185.00 (our price)

Collaboration and Student Engagement in Design Education
Richard Tucker (Deakin University, Australia)
Information Science Reference • copyright 2017 • 411pp • H/C (ISBN: 9781522507260) • US $195.00 (our price)

Handbook of Research on Creative Problem-Solving Skill Development in Higher Education
Chunfang Zhou (Aalborg University, Denmark)
Information Science Reference • copyright 2017 • 500pp • H/C (ISBN: 9781522506430) • US $295.00 (our price)

Handbook of Research on Study Abroad Programs and Outbound Mobility
Donna M. Velliaris (Eynesbury Institute of Business & Technology, Australia) and Deb Coleman-George (University of Adelaide, Australia)
Information Science Reference • copyright 2016 • 898pp • H/C (ISBN: 9781522501695) • US $335.00 (our price)

Handbook of Research on Promoting Cross-Cultural Competence and Social Justice in Teacher Education
Jared Keengwe (University of North Dakota, USA)
Information Science Reference • copyright 2017 • 470pp • H/C (ISBN: 9781522508977) • US $265.00 (our price)

Engaging 21st Century Writers with Social Media
Kendra N. Bryant (Florida A&M University, USA)
Information Science Reference • copyright 2017 • 306pp • H/C (ISBN: 9781522505624) • US $180.00 (our price)

Table of Contents

Detailed Table of Contents

Developing Chinese Design Education (CDE) with Chinese characteristics is an old but unending topic that has been discussed in the field of design education in P. R. China for many years. The debate has been constrained by different attitudes towards the nature of design, diverse interpretations of the Chinese cultural tradition and varied perceptions of the role of education. The future development of CDE is eager for new attitude, thinking and approach to rebuild its culture and value system. This article develops a comprehensive framework through redefining the meanings of CDE, applying the principles of spatio-temporal measurement and ADAL model of Integral Theory. This integral approach based on the newly established framework provides a pragmatic and collective method for CDE, not only from a strategic height to simultaneously understand multiple, competing theories and ideas, but also from tactical operational level to holistically manage different perspectives and practical solutions.

With the impact of the development of the Creative Industries in the UK and other countries, China adopted the term Cultural and Creative Industry in a national strategy to initiate a structural refinement of the industrial sectors. The animation industry in China has long served as a major original equipment manufacturer (that is, producers of contract work for an external brand) to foreign investors. It then started to develop into a center of original design manufacturers (self-originated work sold to others) and original brand manufacturers (self-originated, self-branded work). This led to a rapid demand for creative talent from higher education institutes. Creativity is a relatively complex concept, and successfully fostering creativity in education demands a clear conception of what creativity and creativity education are. The objective of this chapter is to explore how teachers and students perceive the meaning of the term creativity and identify factors that contribute to teachers and students' conception of creativity in education.

In recent years, the number of Chinese students seeking advanced education in the United States has increased considerably. This paper attempts to compare Chinese students' expectations of the MFA program in digital arts in the U.S. to the actual contents offered by the graduate schools. It addresses the unique challenges they face when pursuing MFA degrees and discuss how they could overcome those challenges to successfully achieve their educational goals, as well as how the MFA programs can assist them in this endeavor. It explains that digital arts curriculums in the US are usually meant to help students achieve an individual approach to the field, while developing their aesthetic and artistic sensibilities through conceptualization. But upon entering graduate schools, Chinese students usually lack proper knowledge of the theory, history, and philosophy of art, which are employed extensively in these programs to engage students in a multidisciplinary practice that emphasizes as much on the conception of the final piece as it does on its technical prowess.

The purpose of this paper is to develop further insights into micro-entrepreneurship programs participated in by Chinese industrial designers. A model of creative thinking is employed to explain the campaign creation process. A case study research in sample design entrepreneurs was designed and conducted, and it was composed of three steps: preparing for data collection, collecting the evidences, and analyzing the evidences. It was found that five main defects in creative thinking work as obstacles to crowdfunding success. In order to overcome these drawbacks, it is suggested that designers involved in micro-entrepreneurship programs should acquire the abilities of building prototypes, following the design procedure, finding and solving problems, defining ideation and applying the evaluation methods. Current findings and future study can contribute to the curriculum development for China's industrial design education.

"Industry-academia partnership" refers to the cooperative relationships between universities and enterprises during economic development. The purpose of this study is to establish a cooperative channel, strengthen the two-way exchanges between universities and enterprises, and highlight their complementary relationship. Universities shall guide enterprises toward correct operational philosophy, and assist industrial upgrading and economic development, while enhancing the research level of universities. As a result of this study, industry-academia partnership integrates social resources to give full play to mutual exchange between economic functions and social adaptations in order to promote industrial upgrading.

Designing requires the simultaneous application of diverse modes of thinking and making; the administration of a set of competences, among them creativity, ranked most important by Hong Kong Polytechnic University School of Design faculty members. An introductory design thinking subject offered by the School uses Play as a means to foster best design learning experiences. It leverages Play's integrative cultural agency to facilitate students' acquisition of creative and instrumental thinking skills. From the formulation of a hypothesis and identification of a context, to subsequent iterative development and testing of a design proposition, the acquisition of fundamental creative design thinking skills are facilitated by play and demonstrated over the course of a project. A survey questionnaire (N=219) and subsequent factor analysis revealed positive student feedback. This chapter describes how the subject was implemented, and suggests how this blend of international cultural influences informing design education could serve China's creativity and innovation industries.

Are there lecturers capable of educating a digital generation of players in state-of-the-art innovation and creativity techniques? Far-reaching challenge to accomplish, but

Preface

INTRODUCTION: THE CHINESE 'ECONOMIC MIRACLE'

For the past three decades, China's economic success has been an inspiration for developed and emerging countries alike (Fung & Lo, 2001; Hai & Siu, 2011; Liu, 2016; Siu, 2009, 2003). With the opening reforms, China's massive labour force has attracted companies from around the world and effectively turned the country into the world's manufacturing centre (Hui, 2015; Tischler, 2011; Xin, 2011), bringing investment into the country on a massive scale.

Today however, various factors threaten the country's status as the world's manufacturing centre and consequently its continuous economic success. From increasing labour costs (Hui, 2015; Zhao, 2012) to shrinking global demand for mass-produced goods and a declining number of people of working age (Edwards, 2015), many of the conditions that gave China its competitive advantage over the past three decades are changing. In 2015, after decades of record-breaking performance, the country registered its lowest economic growth in 25 years (Magnier, 2016), with all indicators pointing at a 'new normal' of lower economic growth (Chang, 2015; Economist, 2015; Hui, 2015; Lopez, 2015, 2016; Rapoza, 2015; Richter & Street, 2015; Russo, 2016; Turner, 2016), showing that the economic model upon which the 'Chinese Economic Miracle' was built needs to be updated.

It has been long anticipated that China, along with other emerging nations in the BRICS group, would have to transform its economy from one based on the export of low-cost products to one based on internal consumption and the production of high-value products, which, amongst other things, implies transforming the country into a more entrepreneurial society (Mantzanakis, 2015; Xin, 2011).

The Chinese government has clearly shown its intention and determination to transform China into a society led by innovation by 2020 (Abrami, Kirby, & Warren, 2014; Staats, 2011; Zhao, 2012). Since the opening reforms of President Deng Xiao Ping, the government has financed the development of high-tech parks all over the country, the first of which appeared in Shenzhen, a city known as the 'Silicon Valley of China' since 1985 (Abrami et al., 2014; Wong, 2015). Following the administration of ex-president Hu Jintao, ex-premier Wen Jiabao acknowledged that China needs people like Steve Jobs, capable of creating products that can dominate the world's markets (Zhao, 2012). More recently, President Xi Jinping has also said that China needs to go through three transitions: 'From China's speed to China's quality; from China's products to China's brands; and from "made in China" to "created by China"' (Liu, 2016, p. 53).

The government has also been actively encouraging the creation of start-ups through programmes that encourage entrepreneurship and by offering incentives to Chinese studying abroad, to return and start businesses (Dahl, 2016; Lu, 2015).

The latest in a series of comprehensive plans to achieve this transformation is 'Made in China 2025', a strategy to transform the country's industry by strengthening 10 strategic sectors through the use of technologies and by improving its ability to innovate (cC, 2016; The State Council, 2015; Hui, 2015; Kennedy, 2015).

INNOVATION BY DESIGN

Although it is well acknowledged that the Cultural Revolution prevented modern design movements from emerging in China until the late 1970s (Don, 2014; Liu, 2016; Xin, 2011), design, has been instrumental in China's economic success for the past 30 years. Since the late 1980s, sub-departments of design began to appear in some of today's largest corporations, such as Haier (海尔) and Midea (美的) (Don, 2014).

During the 1990s, as some of these corporations became increasingly aware of the importance of developing their brands, several of them started to transition from OEM to ODM. The trend expanded with China's entrance to the World Trade Organization in 2001, when many Chinese companies realised that they would face increasing competition within the country. This period also witnessed the emergence of some of today's best-known design firms and freelancers. By 2010, China had become the world's factory and many transnationals had relocated their 'design for manufacturing' operations to the mainland (Xin, 2011).

Over the past decade, tier-1 cities in China like Shenzhen, Beijing and Shanghai, were recognised as 'Cities of Design' by UNESCO, and the number of Chinese companies entering for design awards became one of the largest in the world (Don, 2014).

Modern design in China has come a long way since the opening reforms of President Deng Xiao Ping; however, until recently, design only played a supplementary role in value creation, and was often treated simply as a tool for embellishment or even for the imitation of existing products (Don, 2014; Xin, 2011). In contrast, the country's current economic transition demands that design plays a deeper role in value creation, and becomes a driver of radical innovation. Under these circumstances, design needs to transition from design for manufacturing to design for innovation, and to become what Liu (2016) calls 'innovation design', an 'integrated form of innovation for product and service creation in the new knowledge economy', one that can be brought into all kinds of scenarios, 'including engineering, industrial, and service design' (Liu, 2016, p. 53).

From 'Made in China' to 'Designed in China'

To capitalise on this renewed vision of design, several issues must be addressed (Abrami et al., 2014; Don, 2014). While companies such as Midea, Alibaba, Li-Ning, Tencent and Lenovo have made design an essential part of their new product development and business strategies, many others still see design as a secondary activity – an after-thought.

This is a well-known issue amongst designers in the country: top management in many Chinese corporations often lacks a good understanding of the value not only of design, but also of the role that design plays in developing the brand identity of the company. In consequence, design firms and departments often neglect important aspects of the design process that are 'intangible to the customer', such as research, and instead focus on producing large quantities of 'cookie-cut' designs (Fung & Lo, 2001; Justice, 2006; Siu, 2004; Xin, 2011). Even companies regarded as innovative often face criticism due to a lack of originality in their designs (Bell, 2014; Chen, 2014; Wiens, 2016; Wilson, 2015). If China is to become a leading innovation force in the twenty-first century, a new generation of design innovators, one that does not interpret R&D as 'replicate and duplicate', needs to emerge. This cannot but involve Chinese Universities.

DESIGN EDUCATION

If design has been instrumental in China's economic miracle, so has design education. Although it is agreed that modern design education did not re-emerge in the country until the 1980s (Don, 2014; Justice, 2006; Xin, 2011; Yang, 2016), ever since, China has achieved great progress in expanding its capacity to support design through design education (Fung & Lo, 2001; Justice, 2006; Wilson, 2015).

The Chinese government's investment in education over the past two to three decades is unparalleled; more than a thousand new design study programmes have been created (Tischler, 2011), with some sources putting the figure at 2000 (Yang, 2016). The number of industrial design graduates alone exceeds 10,000 per year. With more than 400 colleges and universities offering some type of design degree,

China currently has one of the highest numbers of design schools in the world (Feng & Siu, 2009; Justice, 2006; Sharma, 2011; Siu, 2009, 2000, 2010; Sorabjee, 2012).

Yet, the economic transformation of the country places increasing and new demands on design education (Don, 2014). China needs not just designers, but outstandingly creative individuals. While the well-known inventions of ancient China – the compass, gunpowder, papermaking and printing – are proof of the great capacity of Chinese (Mantzanakis, 2015, p. 11), Staats (2011) also notes that there is a direct relationship between creative progress and the 'cultural climate and controlling power at any particular time in its history' (Staats, 2011, p. 45).

It is this 'climate' in which creativity is expected to flourish that remains a matter of concern, because creativity can be either encouraged or discouraged through the environment (Staats, 2011, p. 46).

Good designers are adventurous, ever ready to challenge old ideas and move in new directions. Such designers do not just happen. They have to be nurtured and educated well in the skills that enable them to think creatively and critically. The training they receive is of vital importance in their later success as creative designers. If design is recognized as a very significant feature of production and services in Chinese commerce and industry, then it will be important to ensure that design education is effective in producing designers with highly creative flair (Fung & Lo, 2001, p. 172).

The Chinese education system is often linked with issues such as a negative effect on critical thinking, creativity and entrepreneurship, and an over-emphasis on scores: 'How can students so completely focused on test scores possibly be innovators?' (Abrami et al., 2014; Fung & Lo, 2001; Xin, 2011; Zhao, 2012). Consequently, there is a common feeling that if the Chinese education system is going to produce creative thinkers, an overhaul is 'urgently needed' (Fung & Lo, 2001, p. 172).

The educational foundations of China need to be shaken and shifted if it is to become a society that generates creativity. The current educational philosophy is a millennia-old product of Confucianism. This system still requires rote memory and absolute submission to ruling authorities, teachers, and administrators. China needs to breakout of this traditional approach to education if it is to progress beyond incremental and imitative innovation to radical innovation (Staats, 2011, p. 50).

THE QUESTION

The historical and cultural dimensions of China make generalisations almost impossible. Depending on the source, the period studied and the angle taken, the country can appear as a creative powerhouse or a creative slum.

While the cultural revolution left the creative capacity of the country severely deteriorated, China's transformation into the world's factory brought big challenges and opportunities for the development of design and design education. These challenges and opportunities were, to a great extent, successfully approached by the Chinese leadership.

Now, China's aim to transition from the world's largest exporter of low-cost products to the world's largest exporter of high-value products is again presenting big challenges and opportunities for the development of design and design education. This time, however, it will require more than big spending on infrastructure; deep reforms are needed to create an environment in which a new generation of highly creative designers can flourish.

The question is, can design education in China create an environment in which not only technical expertise, but also creativity and innovation can flourish? Furthermore, what can other emerging countries learn from China's design education story, past and present and future?

THIS BOOK

This book presents a compendium of contemporary analysis and research on design education in China, the role it has played in supporting the 'Chinese economic miracle' and the challenges it faces in fostering creativity and innovation as the country transitions towards a new economic model based on internal consumption and the creation of high value-added products. It will serve as a valuable reference for anyone interested in applying this experience to the development of design, and design education policies to support other emerging nations moving towards this transition.

In Chapter One, Li Zeng from the University of Central Arkansas discusses the difficulties and challenges that design education faces in China, as the country attempts to transition from being a 'follower' to being a 'leader' of innovation in the world. Based on her experience and research in China and the US, she presents a series of steps to improve the structure of higher education in China, and enhance the capacity of the system to interact with an increasingly multi-cultural global environment.

In Chapter Two, Stefano Ceppi shares his experience as a design lecturer in China as he discusses the importance of having Western experts learn more about the culture of the emerging countries where they work to facilitate knowledge transfer. He discusses the challenge of fostering critical thinking among Chinese students and proposes a new approach, while warning that critical thinking cannot be confined to the classroom.

In Chapter Three, Fang Xu from the University of New South Wales in Australia explores the recent growth of design education in China, and presents the reader with a new framework through which Chinese design education can respond to the challenges posed by the economic transformation of the country.

In Chapter Four, Henry Ma from The Hong Kong Polytechnic University discusses the factors that led to the rapid development of animation education in China, and considers the complications arising from it as education institutions have had to meet an explosive demand for talent. He further discusses the challenges of meeting higher expectations from industry, which demands highly creative, rather than merely technically qualified, graduates.

In Chapter Five, Edmond Salsali from Georgian Court University and Rebecca Ruige Xu from Syracuse University report on a study that offers a glimpse into Chinese design education, reflected through the experiences and performance of graduate Chinese students abroad. They highlight the need to move design education in China beyond technical expertise and discuss the often disparate expectations of students versus what schools actually offer. They also offer a series of suggestions to help design schools prepare their students better for graduate school abroad and ultimately improve their study programmes.

In Chapter Six, Ding Zhou and Jiabei Jiang present a case study from Nanjing University of the Arts, which explores the use of crowdsourcing as a medium to support innovation and entrepreneurship in design education. They assess the lack of skills that this case-study exposes, and the challenges that it poses fo educators and institutions interested in incorporating crowdsourcing into their study programmes.

In Chapter Seven, Jui-Che Tu of the National Yunlin University of Science and Technology and Yu-Chen Huang of the National University of Kaohsiung present a case study showing how industry-academia partnerships are being used to fulfil the government call to advance research in science and technology in Taiwan. They discuss why these partnerships are an excellent way to ensure that research in universities leads to the creation of new and practical technologies and products. The chapter also shows why this cooperation is vital for small and medium enterprises, yet despite being one of the major driving forces of the economy, the necessary resources to sustain R&D programmes are still lacking.

In Chapter Eight, Remi Leclerc demonstrates how, following a unique approach in design education, play is being used at The Hong Kong Polytechnic University to facilitate the acquisition of creative skills amongst product design students. He presents an enlightening reflection of how the ambiguity of play helps designers to deal with uncertainty, an unavoidable variable, particularly in emerging nations. He discusses how the School of Design integrates play into its curriculum as a way to attain better 'design thinking' pedagogy, and the role that Hong Kong can play as a lab where educational innovations can be tested before being applied on the mainland.

In Chapter Nine, Miguel Rivas of the University of Central Lancashire Madrid's campus introduces an emerging paradigm in teaching, as he conceptualises a new type of educator with great potential in design education: 'the neurolecturer'. He walks readers through the characteristics that define this new type of instructor, as well as the challenges and opportunities that this paradigm shift poses to education systems in emerging nations.

Finally, in Chapter Ten, the editors review the emergence of the PhD programme at The Hong Kong Polytechnic University and the strategies through which the programme has met the challenges that afflict doctoral design education. The chapter highlights the importance of setting a design research agenda that can generate knowledge relevant to emerging nations, and the role that PhD programmes can play in it.

REFERENCES

Abrami, R., Kirby, W., & Warren, M. (2014). Why China can't innovate. Retrieved from https://hbr.org/2014/03/why-china-cant-innovate

Bell, K. (2014). 7 things copycat Xiaomi blatantly ripped off from Apple. Retrieved from http://www.cultofandroid.com/66569/xiaomi-dont-copy-apple/

Chang, G. (2015). The Chinese 'century' is already over. Retrieved from The National Interest website: http://nationalinterest.org/feature/the-chinese-century-already-over-12915

Chen, B. (2014, October 10). Apple's Jony Ive has harsh words for Xiaomi. *The New York Times*. Retrieved from http://bits.blogs.nytimes.com/2014/10/10/apples-jony-ive-has-harsh-words-for-xiaomi/

Dahl, J. (2016). Chinese government luring 'sea turtles' home to launch startups. Retrieved from Forbes website: http://www.forbes.com/sites/jordyndahl/2016/05/03/chinese-government-luring-sea-turtles-home-to-launch-startups/#636e061b478a

Don, C. (2014). *From 'Made in China' to 'Designed in China'*. Paper presented at the M+ Matters, Hong Kong. Retrieved from http://www.mplusmatters.hk/importexport/pdf/herenke.pdf

Economist, T. (2015). The economy: Go slow. Retrieved from The Economist website: http://www.economist.com/news/china/21645839-prime-minister-seeks-lower-expectations-economy-go-slow

Edwards, J. (2015). China's massive economic advantage over the world is about to disappear. Retrieved from Business Insider website: http://uk.businessinsider.com/chinas-four-decade-miracle-ends-this-year-2015-8

Feng, W. W., & Siu, K. W. M. (2009). Meeting the challenges of education reform: Curriculum development of technology education in mainland China and Hong Kong. In *International Conference on Technology Education in the Asia Pacific Region Conference Proceedings: Less is More* (pp. 447-457). Taipei: National Taiwan Normal University.

Fung, A. S. Y., & Lo, A. Y. N. (2001). Design education in China: New proposals to address endemic problems. *Journal of Art & Design Education*, *20*(2), 171–179. doi:10.1111/1468-5949.00264

Hai, J., & Siu, K. W. M. (2011). Design for all: Social innovation and service design education and practices in China. In P. L. P. Rau (Ed.), *Internationalization, design and global development* (pp. 514–523). Heidelberg: Springer. doi:10.1007/978-3-642-21660-2_58

HKTDC. (2016). Glossary: "Made in China 2025". Retrieved from Hong Kong Trade Development Council website: http://economists-pick-research.hktdc.com/business-news/article/Glossary/Glossary-Made-in-China-2025/glossary/en/1/1X000000/1X0A5GQY.htm

Hui, L. (2015, June 9). Made in China 2025: How Beijing is revamping its manufacturing sector. *South China Morning Post*. Retrieved from http://www.scmp.com/tech/innovation/article/1818381/made-china-2025-how-beijing-revamping-its-manufacturing-sector

Justice, L. (2006). Product design education in China. Retrieved from IDSA website: http://www.idsa.org/sites/default/files/nec06_justice_lorraine.pdf

Kennedy, S. (2015). Made in China 2025. Retrieved from Center for Strategic and International Studies website: http://csis.org/publication/made-china-2025

Liu, S. X. (2016). Innovation design: Made in China 2025. *Design Management Review*, *27*(1), 52–58.

Lopez, L. (2015). This dazzling visualization shows how everything in China is shrinking. Retrieved from Business Insider website: http://www.businessinsider.com/visualization-china-shrinking-economy-2015-8

Lopez, L. (2016). Old China is drowning. Retrieved from http://finance.yahoo.com/news/old-china-drowning-173858576.html

Lu, R. (2015). China, start-up nation. Retrieved from http://foreignpolicy.com/2015/04/22/china-startup-nation-from-factory-to-tech-incubator/

Magnier, M. (2016, Januray 19). China's economic growth in 2015 is slowest in 25 Years. *The Wall Street Journal*. Retrieved from http://www.wsj.com/articles/china-economic-growth-slows-to-6-9-on-year-in-2015-1453169398

Mantzanakis, S. C. Epaminondas. (2015). *China 2025 research and innovation landscape*. Retrieved from??????? Rapoza, K. (2015). World Bank States The Obvious On China, China Gets Mad. Retrieved from www.forbes.com/sites/kenrapoza/2015/07/19/world-bank-states-the-obvious-on-china-china-gets-mad/#16406d1f1097

Richter, W., & Street, W. (2015). China's economic party is 'running out of booze'. Retrieved from www.businessinsider.com.au/chinas-economic-party-is-running-out-of-booze-2015-5

Russo, J. (2016). Ten reasons China failed: From basic gangsterism to parasitic engagements. Retrieved from Observer website: http://observer.com/2016/02/ten-reasons-china-failed-from-basic-gangsterism-to-parasitic-engagements/

Sharma, Y. (2011). CHINA: Design education for future economic growth. Retrieved from University World News website: http://www.universityworldnews.com/article.php?story=20111118193754828

Siu, K. W. M. (2000). Re-construction of learning space for design education. *Design and Education*, *8*(1), 20–28.

Siu, K. W. M. (2003). Nurturing all-round engineering and product designers. *International Journal of Technology and Design Education*, *13*(3), 243–254. doi:10.1023/A:1026101210221

Siu, K. W. M. (2004). Flexible learning environment for design education: Meeting the changes. In D. Saunders, K. Brosnan, M. Walker, A. Lines, J. Storan, & T. Acland (Eds.), *Learning transformations: Changing learners, changing organisations, changing communities* (pp. 110–113). London: Forum for the Advancement of Continuing Education, University of East London.

Siu, K. W. M. (2009). New trends of research in postgraduate design education in China. *US-China Education Review*, *6*(9), 16–21.

Siu, K. W. M. (2010). Meeting the new needs: Design research education in China. *Research in Higher Education Journal*, *6*(March), 157–173.

Sorabjee, D. (2012). China has 400 plus design schools, much less exist in India. *The Economic Times*. Retrieved from http://economictimes.indiatimes.com/industry/china-has-400-plus-design-schools-much-less-exist-in-india/articleshow/12241365.cms

Staats, L. K. (2011). The cultivation of creativity in the Chinese culture—Past, present, and future. *Journal of Strategic Leadership*, *3*(1), 45–53.

The State Council. (2015). 'Made in China 2025' plan issued Retrieved from The State Council The People's Republic of China website: http://english.gov.cn/policies/latest_releases/2015/05/19/content_281475110703534.htm

Tischler, L. (2011). China knows it must innovate. But can it? Retrieved from Fastco Design website: http://www.fastcodesign.com/1664945/china-knows-it-must-innovate-but-can-it

Turner, M. (2016). Something terrible seems to be happening to the Chinese job market. Retrieved from Business Insider website: http://www.businessinsider.com/china-job-postings-declining-since-november-2016-

Wiens, K. (2016). Huawei just copied the iPhone—Down to the Last Screw. Retrieved from Wired website: http://www.wired.com/2016/05/huawei-iphone-screws-ifixit/

Wilson, M. (2015). Why China is investing In design. Retrieved from Fastco Design website: http://www.fastcodesign.com/3041089/why-china-is-investing-in-design

Wong, S.-L. (2015). Hangzhou shows the way to China's new economy. Retrieved from Reuters website: http://www.reuters.com/article/us-china-economy-hangzhou-idUSKBN0O52NO20150521

Xin, F. (2011). *Industrial design: Contrasting the United States and Chinese methods from the perspective of an industrial designer who has both studied and worked in the U.S. and China. (Master of Design)*. University of Cincinnati.

Yang, I. K. Monique. (2016). Mapping China: Design - Education. Retrieved from Dutch Culture website: http://dutchculture.nl/en/mapping/mapping-china-design-education

Zhao, Y. (2012). lunking innovation and creativity. *Kappan, 49*(1).

Acknowledgment

We would like to thank all the contributors to this book, for the privilege it has been to work with them during the development of this book, and supporting their initiative to share their research and reflections about design education. We also would like to thank the members of the editorial advisory board and additional reviewers, for all the useful leads, comments and help provided during the peer-review process. We thank the editorial team at IGI for trusting in our proposal, and particularly our assisting development editor; Courtney Tychinski, for her support and prompt reply to all of our questions. Similarly, we would like to acknowledge the Public Design Lab of The Hong Kong Polytechnic University for supporting the organization and publication of this book. Special thanks go to our friends and colleagues Cynthia Xiao and Elaine Wong, for reviewing the Chinese translation of the call for chapters that was sent to professors in China, and to our friend and colleague Satyakam Sharma, whose support during the peer-review stage of the process was essential. Finally we would like to thank the Research Grants Council of Hong Kong for the support through the Hong Kong PhD Fellowship Scheme, as well as the support of The Hong Kong Polytechnic University, both of which have been essential in making this book possible.

Kin Wai Michael Siu
The Hong Kong Polytechnic University, Hong Kong

Giovanni J. Contreras Garcia
The Hong Kong Polytechnic University, Hong Kong

Chapter 1
Co-Existence:
Challenges and Opportunities Chinese Design Education Faces

Li Zeng
University of Central Arkansas, USA

ABSTRACT

This paper discusses the design capability challenges that are needed to be overcome by China to transform its current process type manufacturing industries into an innovation-design capabilities facilitated third phase industrialization process. This paper will discuss: 1) China's industrialization process in the context of the relationships between industrialization and innovation and the role that design capabilities play in facilitating the transformation of innovation into globally competitive products; 2) current weaknesses and challenges faced by design education in China and design capabilities needed to be developed to support an industry that is driven by innovation; 3) opportunities and recommendations for future actions.

DOI: 10.4018/978-1-5225-0911-0.ch001

INTRODUCTION

There is a growing awareness in developing countries that their industrialization process will stagnate unless they can leapfrog from a process type manufacturing generated by foreign-owned or foreign-controlled industries into a locally-owned industries innovation-driven-push process. Generally the industrialization process can be categorized into three phases:

1. The beginning of the mechanization of labor during the first phase,
2. Learning to improve the efficiency and effectiveness of the manufacturing process during the second phase and
3. The emphasis on innovation and creativity for the generation of new technology, new processes and new products in the third phase.

Cognizance of the noticeable trends in these phases is significant to our discussions of the role of design capabilities education in supporting an innovation push industrialization process. This paper discusses the design capability challenges that are needed to be overcome by China to transform its current process type manufacturing industries into an innovation-design capabilities facilitated third phase industrialization process. This paper will discuss:

1. China's industrialization process in the context of the relationships between industrialization and innovation and the role that design capabilities play in facilitating the transformation of innovation into globally competitive products;
2. Current weaknesses and challenges faced by design education in China and design capabilities needed to be developed to support an industry that is driven by innovation; and
3. Opportunities and recommendations for future actions.

BACKGROUND

- **China's Industrialization, Innovation, and Design Capabilities:** Bai (2011) referred to China's first phase of industrialization as "processing trade" where the main manufacturing capabilities are limited to processing of imported materials, parts and components, accessories and packaging materials from abroad and exporting the finished product. The attractiveness of China as the "world's biggest workshop" (p.22) at the height of "globalization" is the pull factor of abundant low cost labor supplies and cheap real estate. This phase of industrialization brought a higher standard of living from the huge amount of foreign direct investment, increased exports, employment opportunities and demand for setting up of related industries. However, process manufacturing type of industrialization remained to be low-value added activities. Reliance on manufacturing of outsourced products from companies of developed nations, makes it difficult for China to break away from its dependence on labor-intensive and low added-value driven industrialization process.

The second phase of industrialization for China was brought about by push factors, the forefront of which is the rising labor costs and scarcity of usable real estate. These two factors slowed down the demand for processing type manufacturing facilitated by foreign direct investment. The financial crisis in 2008 further incentivized China to deepen its processing type manufacturing by focusing more on adding value to the final products and increasing efficiency.

As Bai (2011) observed, although overall exports dropped in 2010, increasing efficiency and efficacy of the processing type manufacturing capabilities have contributed to the continued rise of the value added ratio i.e. the ratio of export value to import value. However, in 2015, there is clear evidence that Chinese economy is slowing down, caused by the slowdown in industrial sector growth. As Ligon and Fong (2009) reveal, agencies ought to learn the power of creativity and innovation if intend to survive in the "contemporary inter-global marketplace" and stay competitive. For China to move from the low value-added industrial activities into the competitive world market, there is an urgent need to build its innovation and creative capabilities supported by investments in Research and Development (R&D) and an education system that is focused on science, engineering and technology (Wu, 2011). Innovations which are the results of commercialization of R&D investments have to be supported by design capabilities that are integrated into the manufacturing activities as a strategic business resource.

- **Design and Design Capabilities:** According to Cross (2001), design contributes to the making of a new artifact and its appearance in terms of dimensions, materials, and colors. However, design is not just adding aesthetic attributes to products and services, but the integration of design thinking into every sphere of the creation process. As Cohen (2014) noticed that concept "continually evolving" through his experience with the Stanford.School boot-camp design thinking process: Empathize, Define, Ideate, Prototype and Test. Nag & Noguchi (2003) stated that design process starts with a goal and ends with representations of the goal. In that process, designers must be able to translate an initial goal by understanding users, needs, defining the right problem, brainstorming new ideas, enhancing the aesthetic attributes and increasing the efficiency of production, marketing, distribution and consumption. Cross (2001) also summarizes design definition as cognitively thinking of ways to solve problems by generating and testing potential solutions. Therefore, design integrates people's needs from the utilization of the products or services with the possibilities of the application of current technology to satisfy those needs. Designers have to be able to insert these considerations into every aspect of a product/service planning, design and delivery (Brown, 2011). This means design capabilities must evolve, from emphasis on art and aesthetic appeal that focus on the external appearance or presentation of products and services into a strategic methodology that hatches innovative products/services.

- **Innovation and Design:** According to Rubera and Kirca (2012), innovation in a company setting implies adopting new ideas that lead to the development and launch of new products. However, innovation is not only about creating or developing new products. Atalay, Anafarta & Sarvan (2013) categorized innovation into five types: new products, new production processes, new materials and resources, new markets, and new organizational forms. Innovation is driven by the need to discover or create new markets, develop new methods of production and transportation and awareness of the need to introduce new kinds of consumer goods arising from changing needs. The satisfaction of these needs is through intentional creativity to effect "social or material change" (Sood, 2011). Such concept is similar to what Kolko (2008) suggests, that innovation as "valuable newness" (product, service or system) emphasizes the users' viewpoint rather than the producers' viewpoint. A good example would be Apple designing "new products around the needs of the user, not the demands of the technology" (Economist, 2007). Even non-fortune 500 firms like Telsa also illustrate such users needs in their innovation. Telsa's battery and charging technology made batteries cheaper and recharging quicker. Not only did such new innovation helped Tesla lower costs but also catered its users needs on price and efficiency (Bullis, 2013). In these cases, innovation can be successful only when the "valuable newness" is able to provide the adopter of the product, new kinds of experiences, enable the user to derive pleasure in the utility of the product and perceive quickly the advantage or good in the use of the product.

So, what's the relationship between innovation and design? Kolko (2008) distinguishes innovation from creative efforts. Kolko notes innovation represents business concept and development perspectives while design focuses more on cultural or social responsibilities that involve language, process and humanity. Business perspective focuses on "facilitation, development and management" of new and valuable products, but design can be “affordability, functionalism and purpose”. Innovation may include design as a process of launching products, but it cannot replace design either. According to Kolko, design is not “a small part of any larger discipline”, but “a proper entity of its own”. Friedman (2000) also suggested that design intersects with six very large fields; natural sciences, humanities and liberal arts, social and behavioral sciences, human professions and services, creative and applied arts, and technology and engineering. Barnes’ (2012) stand on designers’ perspective emphasizes that understanding of methods and approaches from outside of the discipline helps designer understand different perspectives. So design is not just the artistic depiction of an innovated product; it should take on the larger role of understanding the different ways in which a product and service is being consumed and how it impacts users and our society. Design can include researching, messaging, prototyping, building, branding, packaging, coloring, selecting materials, presenting, testing, all of which will draw upon a designer’s knowledge of art, culture, science, technology, marketing, management, psychology etc. Hence the challenge to build design capabilities to support innovation have to take into consideration a broader array of skills, abilities and knowledge than that of the artistic aesthetic approach to the training of designers.

Policy makers need to be aware and understand the increased role of design and its relationship with innovation, people and society. Innovation might define certain kinds of successful creativity in the business setting, but design plays a vital role to humanity and human relationships (Kolko, 2008). Chick (2010) also pointed out that social innovation as a trend for the design field and expressed designers’ “desire to meet social needs which can be neglected by traditional forms of private market provision” and was “gaining attention and support from governmental institutions and the third sector (voluntary and not-for-profit)”. The results of social innovation can be any new products, services and organizational forms just like any innovation (Murray et al, 2010). Differing from business innovation, social innovation from designers’ perspectives can also be a principle, an idea, a social movement, an intervention, or some combination of these possibilities (Bjögvinsson et al, 2012; Design Council, 2010). This role of design brings to social, business and technological aspects and can be more critical for countries that are exhausted by the abuse of natural resources, poor quality products and low social trust, and will be adding potential value to of their third phase of industrialization.

R&D has often been touted by economists and international organization consultants, as the means by which a country, that is dependent on process manufacturing for foreign companies, can leapfrog to an internally innovation driven industrialization process. Pan (2009) justifies that design R&D and creativity contributes meaningfully to manufacturing sector and “various governments’ industry planning, business operations management, and particularly in cultural and creative industries.” This is the state in which the success of the third phase of industrialization will have to come from the country's science and technology capabilities, social and cultural responsibilities, specifically the Research and Development and Design creativity efforts to generate innovation.

Most developing countries have invested heavily in science and technology human and infrastructural development, to prepare themselves to springboard into this third phase of industrialization. However, most of these countries have failed to make the connection between innovation and creative activity with design. And although countries like China, have a huge infrastructure support and fiscal policies to encourage the development of design skills, innovation and design seemed to go in different directions (Bai, 2011). Design schools have placed emphasis on recruiting students who are more adept towards the arts. Curriculum in these schools are developed around the enhancements of artistic talents to the extent that form making and copying style still dominates the design education practices. Wang (2009) compared the modern design in China with the west. He indicated that while design in the west has reached the *management design* and *strategic design* level, design in China is still in the *form design* and *professional design* stage.

Understanding of design capabilities in China is still quite shallow, mostly focused on the aesthetic/form role of design rather the pervasive role of design capabilities in all facets of production. Therefore, for China's companies and organizations to make competitive inroads into the global market, China needs to systematically grow designers skills, abilities and knowledge. It is not that China is unaware of the demand for such skills, abilities and knowledge or that the market is not responding to such awareness and increase in demand. In fact, every year, more than 100,000 design students are graduating from over 1000 design schools/university (Zhang, 2011). This trend of increasing demand for design school graduates to fill the need of industry, created many schools or colleges offering design courses. But there are still many challenges and problems hidden in the whole system. As a nation with such a huge population base, many problems the author sees in the design educational system are unavoidable in the current context. Opportunities co-exist with challenges in the design education/industry in China. Opportunities viewed from the Chinese strategic perspective can only exist when challenges are matched with resources and capabilities that could overcome threats from such challenges. The interpretation and orientation of design capabilities development should therefore be steered towards the broader role that design should play.

WEAKNESSES AND CHALLENGES OVERVIEW

Current Weaknesses and Challenges Faced by Design Education in China

Like all developing countries trying to leapfrog the industrialization process, the greatest challenge is to grow an economy based upon local innovative and creative capabilities. In the perspective of the three stages of industrialization that were mentioned earlier, innovation can similarly be categorized into three forms which are distinguishable and associated with the three phases of industrialization. The first relates to a local incremental innovation made to improve on technologies adopted from without. In economics, it can be categorized as improvement arising from learning to be more efficient, by making small changes to adopted technologies.

These changes can be a result of local attempts at making the technology more compatible to the local skills, abilities and knowledge to use them. Words like copying, imitating and reverse engineering have been used to describe the innovative activities at this stage of the industrialization process. At this stage of the process the developing country is struggling with trying to integrate the indigenous R and D efforts into the process manufacturing type of industrialization (Wu, 2011). Intellectual property rights usually took a backseat as cheap imitation and copies of protected intellectual properties flood the local market.

The second wave of innovation comes into play when local human resources are more familiar with the technology and as the human resources that support this industrialization process become more knowledgeable and educated, they begin to understand the know-why of the technology being used, thus they are able to make significant changes to those adopted technologies thus contributing towards significant and visible effects on the building up of competitive capabilities. At this stage, products are innovated and local contribution to this innovation is evident by increasing competitive advantage. The intellectual property legal infrastructure begins to take shape with the formulation of laws that adhere to international standards.

The third phase of innovation is the active role played by design in the innovation and production process including contribution of globally competitive products that have achieved global design acceptance status. Both design capabilities of the product and the production technologies adopted to produce the product should have achieved world significance. A good example is South Korea and the many internationally successful brands for household products, automobiles and wireless phones. It is difficult for developing countries to breakthrough from the second stage industrialization process to the third stage, unless the third stage innovation process comes into existence.

Our evaluation of the Chinese entrepreneurs seemed to indicate that they have gone beyond the stage of copying of existing products of developed countries. In most cases the basic design of the products, trademarks and brand names are established and there does seem to be a role for the knowledge generated by R&D. In contrast, during the first phase of industrialization, the science and technology capabilities are not in-sync with the science and technology needs of the industry.

In the beginning the market is growing and there is huge demand for products which have designs, trademarks and brand names that are known to consumers. There is little motivation for the entrepreneur to court unnecessary risks by investing in design of a new product with new aesthetic attributes as there is a lack of competition for established products or service. Thus at this stage, Chinese designers and companies are known to prefer to copy or imitate known ideas and products especially when the legal infrastructure to protect intellectual property rights are still in its infancy. There is also the reservation among policy makers to commit to the enforcement of intellectual property rights and thus stymies innovation. This often happens when the complexity of intellectual property claims and counterclaims are seen by policy makers as a critical factor that could stymy innovation and creativity therefore jeopardize the main goal of stimulating industrial sector growth. It is a nationwide issue that all level of individuals or organizations face.

So, why is it so difficult for countries like China to progress to stage three innovation that could support a stage three industrialization? There are numerous challenges that China has to face to overcome. Wang (2009) listed eight major forces that drives the future direction of Chinese design education: evolution of China's new industrialization; global economic integration; internationalization of education; digital media technology; the internet technology; cultural and creative industries; growth of service industries and lastly cyclical economy. In this chapter, the author will discuss infrastructural and fiscal policy problems, the political will, the fundamental culture beliefs and foremost of all the educational challenges.

EDUCATIONAL CHALLENGES

1. **The Constraint of Rote Learning:** The negative aspect of rote learning and exam oriented education system is that it influences students to focus on methods that could enable them to regurgitate knowledge and capabilities to memorize whatever their teacher or professors disseminate to them. This behavior is prevalent throughout the traditional rote learning system upon which the modern Chinese education system is based upon. The same system is applied to science and technology education as well as art and design education. Design students face a dilemma: where their field depends on their ability to think

and express creatively and individualistically; but success in the exam system depends on them to stay on course on what was taught. The curriculum for design courses are always packed tight with subject matters which might not complement the emergence of creative thinking and innovation and students have to continuously prepare themselves with numerous exams that test their mastery of knowledge transmitted by their professors or teachers. There is no time, energy or even patience left for students to understand the "know why" or the "know how" of the knowledge delivered to them.

2. **Pre-College Art and Design Education:** Rote learning occurs most commonly on a variety of subjects before or during students' high school years. For students who decided to pursue a creative journey since young, many were sent to technical art classes on weekends that was similar to rote learning. Despite the types of medium involved, most of the works/topics students are being taught are simply realistic style of figures, faces, landscapes and still objects, thus few courses that targeted on training creativity and enhancing designing & hand-making skills was involved. Such technique-focused training was more intensified for students who have a creative impulse and strong ambition to attend design schools in college because China's national examination systems helps select talented students across the country. However, because the art/design education Chinese students received prior high school does not match the admission requirements from college level. Many private training studios that target on improving students' art/design ability thrive due to the big needs in the market. Such training purposely focusing on technical development during the final high-school year of art/design students, inhibit creativity overall. The following section will reveal what such pre-college training process is and how it has been hampering creativity for Chinese design students.

Students who want to be admitted by top design schools, like most art students, in their last year of high-school, would devote five to six months full time for more intensive training on drawing and painting in a studio. Then by the last spring semester in February (4 months before the national entrance exam in June), students have to take the national art admission exam hosted by the specific school at a local test center. Each one of these tests consists of 3-4 subjects and each subject were required to be completed in three hours. Preparatory (考前班 Kaoqianban) training studios (画室 Huashi)established by top design school graduates are favored by students and their parents to prepare for the tests. Students also often leave their hometown and find a training studio in metropolitan areas like Beijing, Shanghai or Hangzhou where some of the leading art and design academies locates.

Why does such intensive training matter at all? From the interview with several students in a training studio called "New Generation" in Changsha, Hunan province, it was indicative of a disaster for students because either they have no idea what will be tested, or their technical skills has not reached a level that is competitive enough to pass the art admission test (New Generation, Interview, 2015). Because there are specific categories (figure drawing, still color object, quick sketch) regulated by most universities and colleges, students define their artistic/creative strength and weakness based on the technical levels of those categories. An extreme example would be: some studios teach students how to paint an apple with 10 strokes. This training leads most students to an understanding of creativity/concepts as a freedom to use different techniques (i.e. self-portrait, mosaic style, etc.) that universities didn't list on their admission requirements (New Generation, Interview, 2015). Nahavandi (2015) demonstrated that "visualization" is only one aspect within the eleven elements that judge ones' creativity. The author found that most design institutes' admission tests, though intended to admit the best creative talents, ended up pushing students away from being creative.

The growth of these types of training studios convinced us changes in the design admission process are necessary for evaluating and admitting a design talent. Hu Qiong, founder of "Success Track", a training studio in Beijing, expressed that there is a big hole in the current high school art/design education. She addressed that 80% of the training students received in the studio is focused on techniques. However, the training studio is merely a "shadow" industry for design schools, because they only reflect and respond to what the higher education institution requires. Whatever changes an art institute makes for their admission test, the training studios will incorporate into their students' training (Hu, Interview, 2015).

This survey of art education in preparation for college showed the author that China has a significant challenge in order to cultivate students' creativity prior to college education. Creativity needs to cultivated over a long period and cannot be accomplished in a short period of time such as that offered by training studios. Substantial efforts to promote creativity are necessary throughout a student's education, starting at a younger age . As Staats point out, the cultivation of creativity should start early in a person's life and the environment, families, and schools all contribute to the development of creativity. (Staats 2011) Because the creative education Chinese design students receive prior to college is years of responding to a rigid exam system, the development of a creative mind that can generate personal ideas or innovative thinking contributing to the development of China's innovative industrialization, fails to root in a Chinese design student in his or her formative years. These pre-conceived technically-focused knowledge of design and perception of a designers role on solving simple aesthetic issues may even negatively influenced design students many years after they have graduated from college. Rethinking of ways of connecting the needs of design talents before college creative education and improving university's admission process of design talents is imperative.

3. In College Design Education: As the author's study shows many colleges admitting design students, vary in their standards of skills, abilities and knowledge of design graduates. Consequently, the quality of students' ability is defined based on their choice of higher education institutions. Pan (2009) mentioned that the "role of ambitious administrators bent on expanding the size of their schools" affected the design education negatively. However, even with the best students selected, great resources allocated in the top design schools in China, there are still some issues worth examining. Due to the leading role some Fine Art Academies are taking for training future creative thinker and innovators, a closer look of a representative graduates from one specific school- Central Academy of Fine Art (CAFA) will be addressed as a case study in the following section.

CAFA is one of the 8 leading art and design institutions teaching design in China and has been named as one of the 30 best design school in the world (Business Week, 2009). Wang Min, Dean of the School of Design at CAFA, noted there is a gap between the training for creative thinkers and the understanding/need from the market place. This gap existed because of the difference between the public understanding and expectations of design and the students' interpretation and creation of what the market needs.

Whereas the art minded design students focus on conceptual ideas that might not be practical, they blame the consumers or the market when those ideas do not work. They do not understand that their conceptual ideas should extend into responsibilities of execution and studying of what the market actually wants or willing to accept from the design of the products or services. Students from schools like CAFA should be trained not only in conceptual ideas but should be constantly reminded that the purpose of design is to serve its utility and consumption and therefore should be provided with adequate exposure to social experiences that could make their design skills training more applicable.

Yu Zhou, an alumna of CAFA, for example who pursued a postgraduate degree in England. Currently she is the founder of YUZZO, a successful fashion brand in China. Zhou prefered design education in England due to the opportunities for internship offered by her Britain schools where she was given opportunities to work with "big brand names" in her second year. These companies have their core creative teams, therefore working with them had given her the knowledge and skills needed to start her own brand. From these practical experiences, she developed the skills to understand the needs and create designs that could really contribute towards the utility and consumption of the products and services. (Zhou, Interview, 2015)

Jing Tao, alumna of CAFA, who studied in Central Saint Martin academy in England for her graduate school, and is currently the founder of "Useful" product design brand in Beijing have the same experience. Tao claimed that Chinese design schools lack the curriculum that could provide students with the practical experience to discover the market, which is critical for product success and apply design capabilities within the society. Many students' perspective was restricted by what was taught in the curriculum and what was required in the evaluation process, leaving a gap between designer's own goal and the market needs. (Tao, Interview, 2015)

Jingling Li, alumna from CAFA and now an interactive designer for Wechat in Guangzhou also indicates that fine art colleges kept design students in single academic environment. Students study hard and worked diligently but no social experience was involved at all. Li also pointed out that compare to design professors in the west, Chinese design educators themselves, lacked sufficient professional experiences outside of academic settings to begin with. (Li, Interview, 2015)

These examples confirm what has been pointed out by Bombay (IIT, 2009) who defined modern design as a "multidisciplinary activity that includes professions besides engineering". There is a common narrow interpretation of what design involves in the curriculum for training of designers in higher education institutions. Most interpreted design capabilities as art biased, therefore, the design school is always located within the art department in Chinese universities. Chinese design education curriculum seemed to be narrow and therefore is characterized by limitation in course selections and lack of the choices that could enable the pursuit of a multi-disciplinary knowledge that holistic design education needs.

Zhang claimed that the industrial design major in a Chinese university has two main types: "engineering based design" and "art based design". He pointed out that there is not much cross-discipline interaction between "engineering based design" and "art based design" (Zhang 2011). Many schools that offer the industrial design majors, simply teach pure science courses or basic art courses, there is not a good system that overlaps knowledges or interacts with both fields. Not mentioning that most schools set up their design programs only on an aesthetic level and students themselves are not equipped to integrate these skills and knowledges learned from both fields. The result of such separated knowledge in design is that students in engineering schools produce technical products that doesn't look aesthetically pleasing, but students in art schools can only create products ideas that simply cannot be implemented. With more and more industries overlapping with design, the designer's role increases its importance, multidisciplinary knowledge beyond aesthetics will be more and more vital for a designer to possess when producing creative but practical and implementable innovations in a fast growing economy.

Creative Culture Challenge

The constraint that has often been cited as a major challenge to innovation and creative thinking, is the entrenchment of Chinese traditional tenets within the psyche of the Chinese population. As a country that has more than five thousand years of history, traditional Chinese culture has affected every aspect of the Chinese way of life.

1. **Respect for the Status Quo:** Chief among the impacts of traditional thinking is respect for the status quo. For example one of the commonly cited tenets is that the younger generations should not question the actions and decisions of the elders. There was indeed a tendency in traditional Chinese culture to belittle the technological innovations. A convincing example would be Zhuang Zi's (《庄子·外篇·天地》) reservation about the invention of the well crank. In the story, Zigong was offering a local worker a creative technology to improve his efficiency on farming. While being aware of the technology's advancement, the farmer refused to adopt it due to rote learning and respect of his teacher who claims that people who crave for technology are morally wrong and spiritually not peaceful. The modern definition can be extended to the reverence given to authority and actions and decisions of the establishments among which are government organizations and education institutions. Patience is seen to be a virtue and upsetting the status quo can result in being ostracized or shunned which could be traumatic in a highly collective culture. Within the context of such a social norm, it is difficult to cultivate creativity or innovation. As Staats (2011) pointed out, following the tenets can have a strong counter-productive influence on the development of creativity in China.
2. **Lack of Critical Thinking:** Also, critical thinking is not cultivated when students are discouraged from questioning or from analytical thinking, but, instead are asked to accept what they are told to learn. "Teachers typically do not allow their authority to be challenged. Students are afraid of their teachers, and dare not ask "provocative" questions" (Ho & Ho 2008). Their performance is measured by how well they score on the exams and, academic achievement is prized. Within the context of a culture that prizes academic achievement above creativity and innovation, performance in exams are critical to a successful career. Therefore there are exams at each critical stage of the Chinese students' academic life. The most important of which is the National Exam at the twelfth year of the student's academic life. The students twelve years of academic training culminate into how well they are going to do in the national exam that will decide for them what future they have for the rest of their life. It is not surprising that with this mindset, students will resort to all means to stay ahead of their peers. This emphasis on academic achievement might lead

to dishonest practices such as fraud and cheating on tests and the writing of papers. (Huibert 2005) Huibert observed that such attitude is destroying Chinese students' ability to think independently and creatively. Younger generations who are the output of such a system tend to be more accepting rather than questioning, following rather than leading and focusing on short term gains and quick solutions rather than going after the essence of what an education will enable them to do and that is ability to think creatively and solve problems.

3. **Uncertainty Avoidance:** One cultural dimension that is particular to Asian culture, especially those cultures that are not adept at managing risk and uncertainty is called uncertainty avoidance. Uncertainty avoidance are evidently practiced in societies that emphasized the appreciation of status quo and treasures harmony not conflict, therefore shuns initiative to effect changes especially measures that could inevitably lead to modernization of societies. The traditional mindset that is prevalent in Chinese society to a greater extent is reflected by these characteristics of a high uncertainty avoidance culture. The fear of uncertainty is a psychological state that springs from a mindset that being wrong is an unacceptable outcome. Therefore the probability of being wrong decreases as uncertainty is reduced.
4. **"Face" Culture:** The constant need to be right has given rise to an attitude called the retention of "face" in the Chinese culture. "Face", the giving and taking away of which is extremely important to our discussions of constraints that could hamper the development of creativity, innovation and design capabilities. "Face" has been attributed to China's adoption of a high uncertainty avoidance culture. To lose face is to lose credibility and pride which will result in the erosion of a person's social standing in a collective culture. As Staats (2011) observed, it is unlikely that members of a high uncertainty avoidance culture will take the risks that could result in losing face to venture into discoveries activities or subscribe to activities that could develop new or creative ideas. Failure of such ventures could lead to embarrassment to oneself and family. Therefore in the Chinese society, pragmatism would mean targeting activities which have higher probability of success and to steer away from activities that are new or what is termed as breaking out of the box. This mindset does not fit with the mindset that is required of an innovative and creative culture which requires entrepreneurial spirit and the acceptance that failures are the foundation to success.
5. **Collective Culture:** The collective culture-typology identified by Hofstede (2015) can also be used to explain why new ideas are not forthcoming and creativity is usually stifled in China. Central to the collective culture is the need to conform and Chinese students especially are susceptible to the power of the group used to ensure conformity. They seldom want to deviate from group

norms, at the least they will be chastised as disrespectful, not filial, disloyal, or impatient. At the worst they will be sanctioned and sometimes excised from the group. Thus, Wei and Li (2013) observed that in a high collectivism culture, individuals in a group focused on maintaining group harmony by being attentive to group norms. In the Chinese society, harmony and hierarchy is translated into norms in which those who are low on the hierarchy scale are not expected to play a decisive role. Their ideas are often deemphasized by older and hierarchically more senior members of the group. These lower hierarchy members tend to be less assertive in supporting their ideas. Although sometimes their ideas are better, these ideas have to make way for mediocre ideas made from more senior members of the group (Wei & Li, 2013).

Environmental Challenge

The last major issue that will be discussed that China will have to overcome is the inequitable reward that professional designers are experiencing from the market compared to the reward of designers from countries in the west. This is a common phenomenon in most developing countries and China is no exception. According to a ranking published by *Standard Ranking Research Institute* in 2015, The average salary for a professional designer who graduated after five years in China is around ¥7,110-8,532/month ($1,146-1,376/month), whereas an interactive designer in the U.S. is around $4,583/month, and this number can reach up to $8,333/month if one is promoted to be a creative design director (AIGA, 2014). The result of this inequity is the phenomenon known as brain drain.

1. **Brian Drain:** Brain drain traditionally has been associated with physical movement of valuable human resources from developing countries to migrate to developed countries. However, in this modern context of information technology, especially in the field such as design, the actual work can be done in China itself although the contracted work is in other countries. This made it easier for brain drain to occur as foreign design consulting companies can source Chinese designers for a fraction of the cost to work on their projects. The attraction to local designers who make the grade to work with such companies is the higher rewards they can obtain from such companies compared to working with local companies. Usually these designers are the top of their profession. An indication of this phenomena is the preference given by professional designers in China for designing jobs from overseas market. Chinese designers promote their skills to these international companies by actively participating in international competition. This is the platform that they use to actively solicit for overseas jobs. Therefore, such an occurrence deprived

the local design needs of the much needed professional design services of the nation's top designers. In the macro economic context, such design skills that are needed to bridge the innovation and creative effort of the country with the market are lost. This can cause set-backs to China's aspiration to leapfrog into an industrialization process that is capable of producing world class product designs.

2. **Low Demands in Design:** Another challenge the environment reflected is the designer's role in the current Chinese market. The rapidly expanding manufacturing sector in China meeting the dynamic industrialization process, has generated a huge demand for designing skills. However, this demand reflects a process manufacturing industrialization process, therefore the majority of the demand is focused on design capabilities to support currently successful products and enhancing the aesthetic appeal of the product attributes. The concerns of designers are not on understanding the role of design in commercializing or even generating innovation. The major demand is coming from the lower part of the manufacturing industry that is more generalized and more end-product design rather than generation of new products or processes. In contrast, U.S. designers faced with a dynamic environment, respond to changing market demand by focusing on refinements of brands, draw attention to distinctiveness of products and services and utilization of intellectual property rights to protect the uniqueness of their design (Fan 2011). This short term orientation of Chinese designers to go for instant gratification prevent them from developing skills that could support core technical breakthroughs that may take a longer and harder process.
3. **Government Involvement:** The last factor that deprives good design capabilities from the Chinese market is caused by the ever confusing government hierarchy and jurisdiction over the education and development of design capabilities. The design capabilities development program seems to be stifled by over regulation. Government programs to develop design expertise were not driven by market demand but rather by what the government observed to be important. For example, all programs including education and design education was driven by the Chinese government declaration that China should be transform into an "innovative society" by 2020 and a world leader in science and technology by 2050 (Abrami, Kirby, & McFarlan, 2014).

These overarching goals set the tempo for government policies and incentives and both local government officials and common citizens are told to observe this strategy. An example is in the relationship between educational institutions and the government who administer such academic institutions. A Secretary of Chinese Communist Party Branch was set up to decide on the policy issues and direction of development of education programs in the schools (Yu, Interview, 2015). However, the Presidents of the schools, were given the responsibilities of managing the overall academic affairs of the schools. Such system provides a tool for implementations of governmental policy thus limits the natural growth of certain areas in the institutions. As Ho & Ho (2008) observed that a tightly controlled society does not foster "rugged economic individuals" or "innovative entrepreneurs". Hence loosening control is a precondition for fostering creativity—an assertion supported by considerable evidence (Ho & Ho 2008).

SOLUTIONS AND RECOMMENDATIONS

Future: Opportunity and Recommendations for Design Education in China

After our discussion of the various constraints that are hampering the development of China's design capabilities, it must be noted that the government's continued commitment to investing in design education has had an overall positive impact. There is a positive trend of heightened demand for design capabilities to complement the many innovations that are generated by massive inflow of government funds. Further demand for design capabilities also stems from the booming Chinese economy, which has created a revolution in design and innovation among the new generation.

- **Driven Young Generations:** Generations born after the '80s are considered to be the driving force of the new age of Chinese design and innovation. Lorraine Justice (*China's Design Revolution* 2012) refers to them as China's "third generation" because it is post-Revolution and post–Cultural Revolution. The third generation came of age during China's "opening up" period, and that generation is known to strive for fame, fortune, and self-expression. They are now in their thirties and, perhaps, forties and have had much more freedom to create—and to consume—than their parents or grandparents.
  - Influenced by Mao and Confucius, communism and capitalism, patriotism and cosmopolitanism, the third generation will drive the culture of design and innovation in China—and perhaps the rest of the world. This influence is only strengthened by the government's massive investments in design and innovation and its support of design and creative activities at the local and national levels. Projects such as design education programs, innovation parks, and privatized companies are all aimed at stimulating economic growth and establishing China as a global creative power.

  - The third generation's perception of traditional values and cultural beliefs has been and will continue to be modified by their experiences during the rapid changes to China's economy and society as a whole. This more open and diverse environment allows students to become more passionate and motivated to create value for social enterprises.
  - According to Abrami, Kirby, & McFarlan, 23.9 million students attended institutions of higher learning in China in 2012— 4 million more than were enrolled in the U.S.. (Abrami, Kirby, & McFarlan, 2014) With so many Chinese students attending universities and colleges every year, including those who choose to study abroad, design students are brought into a more open, diverse and dynamic environment. When open environments, abundant resources, quality teaching and overall better college design educational experiences are brought together, there are many opportunities for quantitative technique training to lead to qualitative transformations for Chinese students.
  - Those students who are admitted into top design schools are ensured to possess excellent drawing, sketching, painting and crafting skills. The technique-heavy admission test aligns with the traditional understanding of 'creation' that has persisted since ancient China. Since the time of China's dynasties, artists who created great pieces typically did so by intensively studying or copying various artists from earlier dynasties. Under this model, qualitative transformation occurs after a massive amount of practice has been conducted over the course of many years. Thus, students are expected to intensely train techniques before college, so that they can then be appropriately exposed to new design concepts while in school. The hope is that Chinese design students are able to ultimately reach a point where they can understand the conceptual and creative process while also being able to execute their concepts.
- **Environment- Digital Age and Cross-Cultural Experience:** According to McKinsey & Company's research, which covered 630 million Internet users across different city tiers and rural areas,"Chinese e-consumers are embracing and demanding more innovative digital experience overall". People claimed willingness to try different services or products and displayed an openness for new products and digital age innovations.
  - Such openness is vital for the acceptance of new and creative design products. Even for rural dwellers, 85% of those surveyed used mobile devices to go online (McKinsey & Co 2015). New concepts no longer have geographical barriers due to the rapid development of digital devices and accelerated online experiences. Young social entrepreneurs

who understand current market needs have started to take the lead in digital innovation for society.

- Somewhat related to the new opportunities created by the digital revolution is the opportunity for multi-cultural experiences. In higher education, many design schools have started to collaborate with international institutions by establishing dual-degree programs and exchanges involving students and professors. For instance, CAFA collaborates with Peking University and Stanford University on cross-cultural projects that require cultural understanding and innovation based on contexts; CAA (China Academy of Arts) has established a summer program with Minneapolis College of Art and Design (MCAD) that allows students to participate in workshops in a different culture and environment; and Shandong Art and Design Craft School has been receiving national funds to send design professors to Northeastern University in Boston for research and study purposes. Such efforts open horizons for students and give them experience that cannot be received merely from the domestic environment.
- The thousands of Chinese students that study abroad typically return home with great vision and new ideas for future development. In fact, the Chinese government encourages these returning talents through programs such as the *Hundred Talent Program* and the *Cheung Kong Scholars Program* (Wu 2010). Local governments, like Beijing, also offer advantages to returnees when applying for local residency, which can be quite difficult and time-consuming otherwise.
- As Wu states, such efforts are an encouraging development and they tend to increase academic quality and the diversity of ideas. The more one is open and exposed to different environment, the more one can create and connect to different experiences. An open mindset and acceptance of foreign cultural influences greatly benefits Chinese innovators and creators because diversity is educational and a source of inspiration for ideas that can stand on their own in local contexts.

- **Reflections and Innovation in Education**: According to Abrami, Kirby, & McFarlan (2014), Chinese began to adopt American academic mantra "that the best leaders are those with the broadest education in the liberal arts." The goal of a liberal education is not to train specialists but to educate students to be curious, thoughtful, and skeptical. Many top institutions have begun to understand the deficiencies of a massive training and purely technical focused knowledge delivery system.
  - Reflections from all levels demonstrate that changes are underway. Strong efforts are being put into improving the current education sys-

tem through innovation, and the nationwide "211" program is one such effort. The 211 program was established in 1995 to improve overall capacity and develop key disciplines in science and technical schools through increased funding. Although many of these schools focused on science rather than humanity programs, the overall improvement to the school's quality affected the development of the entire university's academic atmosphere.

- Unlike in the past, Chinese students in leading universities are often encouraged to take courses outside of their majors, and many schools have even started allowing students to change majors and receive multi-disciplinary knowledge and training. However, changing majors is still something that is very difficult to do in Chinese universities. Although this might seem a given in American universities, Fudan University's decision to allow students to "declare a major during their second year despite the direction they have chosen during college-entrance exams" was seen as a somewhat radical move. (Wu 2010). General improvements to colleges such as these may not spread fast enough to cover design schools in every university, but many leading design institutions have been seeking a variety of approaches to establish new programs and majors that allow multi-disciplinary interactions.
- There are also improvements in the existing college design admission test, elite design schools are attempting a variety of ways to involve creativity in the exam system. According to Hu Qiong, one of CAFA's past admission tests required applicants to convey the feeling of tasting a lollipop through illustration in 3 hours. During the test, every student was given a lollipop so that they could have the real experience while creating. The test required students to use all five senses of sight, smell, touch, taste, and feeling when preparing their drawing or design (Hu, Interview, 2015). Unlike some school's admission tests, which require students to draw an exact model from a photo, CAFA's test challenges students to think beyond tangible imitations, and it is a great step toward improving the current creative exam system.
- It is not that policy makers do not realize that current problems exist in our education system and test-oriented structure; fundamental change requires monumental effort in any culture, particularly in one with an enormous population base like China's. Patience is key and small improvements are being made constantly. The impact may not be massive, but these aggregated attempts are the beginning of a new future.

FUTURE RESEARCH DIRECTIONS

There is a growing awareness and social responsibility derived from Chinese designers and educators wanting to turn the "Made in China" label into a "Design in China" one.

The development and growth of the design industry needs domestic buying power and support from the massive Chinese market. So not only do designers have to be devoted to creating new valuable products, but the producers must also take part in the chain in manufacturing, packaging, branding and advertising in the marketplace – this is the way to to empower products rather than diminishes their quality.

If there was a perfect design education system that would train and make perfect Chinese designers, then these designers could come up with creative and valuable products. What would happen if those products were put into the marketplace to compete with internationally-known brands and products? Would people support the domestic designed products and brands? Or would there be more trust toward traditional Western brands? More research can be conducted to determine whether Chinese consumers want quality brands or do they want products that show their social status (Or some other reasons). Designers will be able to produce meaningful services and products that can really sell only when they understand the aspect of business, culture and societal needs.

Another research opportunity from this chapter would be social design entrepreneurship. This social design entrepreneurship trend has become popular in Western countries. China should develop this same trend, too – one that focuses on directing business entrepreneurs to serve the different groups that contribute to good causes and solve social and local problems. Even with tiny but collective efforts made through different levels of designers and groups (in fact, there are many designers and groups are already doing so), the overall design culture in China should be dramatically different. Such research will help motivate the design entrepreneurs at the same time setting models for design students who are eager and passionate to utilize their skills and make a difference to their country.

CONCLUSION

All disciplines face their own educational problems in different contexts, and Chinese design education has several identifiable hurdles to overcome. One cannot point to cultural perception, systemic challenges, or environmental threats as being the single cause of China's existing problems. There are no quick-and-easy solutions to the issues discussed in this paper, but there are a plethora of modifications that can be made to alleviate these problems. As discussed, there are a variety of improvements available for setting up higher educational structure, adapting to the new global market and interacting with multi-culture environment.

The growth of the creative minds of the new era and future generations is of paramount importance. Keeping that in mind, with some more patience, a better future will come for Chinese design education once the nation's overall perception and respect for design continues to grow. Like the growth of Chinese e-consumers, the author also believes that Chinese design education would soon reach "its rapid transition from follower to leader in the world" (McKinsey & Co, 2015).

REFERENCES

Abrami, R. M., Kirby, W. C., & McFarlan, F. W. (2014). *Why China Cannot Innovate?* Retrieved from https://hbr.org/2014/03/why-china-cant-innovate

AIGA (American Institute of Graphic Arts). (2014). *2014 national median total cash compensation*. Retrieved from http://designsalaries.aiga.org/#salaries-list/

Atalay, M., Anafarta, N., & Sarvan, F. (2013, April). The Relationship between Innovation and Firm Performance: An Empirical Evidence from Turkish Automotive Supplier Industry. *Procedia: Social and Behavioral Sciences*, *75*(3), 226–235. doi:10.1016/j.sbspro.2013.04.026

Bai, M. (2010). *The Future of "Made in China"*. Retrieved from http://www.chinatoday.com.cn/ctenglish/se/txt/2011-12/30/content_417573.htm China Today.

Bjögvinsson, E., Ehn, P., & Hillgren, P. (2012). Design things and design thinking: Contemporary participatory design challenges. *Design Issues*, *28*(3), 101–116. doi:10.1162/DESI_a_00165

Brown, T. (2008). Design Thinking. *Harvard Business Review*, 84–96. PMID:18605031

Bullis, K. (2013). *How Tesla Is Driving Electric Car Innovation*. Retrieved from https://www.technologyreview.com/s/516961/how-tesla-is-driving-electric-car-innovation/

Chen, J. L. Zhang, F. X (2011). A Study of S&T Investment Strength and Development Tendency in Foreign Countries. *International Journal of Business Administration, 2*(2).

Chick, A. (2012). Design For Social Innocation: Emerging Principles And Approaches. *Iridescent-Icograda Journal of Design Research, 2*(1), 56-60.

Cohen, R. (2014). *Design Thinking: A Unified Framework For Innovation*. Retrieved from http://www.forbes.com/sites/reuvencohen/2014/03/31/design-thinking-a-unified-framework-for-innovation/#6fa6dcf856fc

Cross, N. (2001). Designerly ways of knowing: Design discipline versus design science. *Design Issues*, *17*(3), 49–55. doi:10.1162/074793601750357196

Economist. (2007). *Lessons from Apple: What other companies can learn from California's master of innovation.* Retrieved from http://www.economist.com/node/9302662

Ho, D. Y. F., & Ho, R. T. H. (2008). Knowledge is a Dangerous Thing: Authority Relations, Ideological Conservatism, and Creativity in Confucian-Heritage Cultures. *Journal for the Theory of Social Behaviour, 38*(1).

Hofstede, G. (2006). Dimensions of Culture. *Journal of Cross-Cultural Psychology, 15*(4), 417-433.

Hsiao, S. W., & Chou, J. R. (2004). A creativity-based design process for innovative product design. *International Journal of Industrial Ergonomics, 34*(5), 421–443. doi:10.1016/j.ergon.2004.05.005

Huibert, D. M. (2005). *Teaching Management to Chinese Students: a few notes from theory and practice*. Amsterdam: University of the Netherlands.

Industrial Design Centre. (2009). *Design as a strategy for developing economy*. Bombay, India: Indian Institute of Technology.

International Council of Design. (2009). *Central Academy of Fine Arts recognized by BusinessWeek*. Retrieved from http://www.ico-d.org/connect/index/post/1365.php

Kolko, J. (2008). The Tenuous Relationship Between Design and Innovation. *Routledge's Artifact, 1*(3), 198-203.

Li, W. H., Rosanna, S., & Angelina, S. (2007). *Creative Industries: Singapore and Hong Kong – A Review of Design Initiatives with Implications for the Nurturing of Design Talent*. Retrieved from http://www.ln.edu.hk/mcsln/20th_issue/feature_01.shtml

Lorraine, J. (2012). *China's Design Revolution*. Cambridge, MA: The MIT Press.

McCraw, T. K. (2010). *Prophet of Innovation: Joseph Schumpeter and Creative Destruction*. Belknap Press. doi:10.4159/9780674040779

Mckinsey & Company. (2015). *China's iConsumer 2015: A growing appetite for choice and change*. Retrieved from www.iberchina.org/files/Chinas-iConsumer.pdf

Murray, R., Caulier-Grice, J., & Mulgan, G. (2010). *The open book of social innovation*. London, UK: The Young Foundation & NESTA.

Nagai, Y., & Noguchi, H. (2003, December). An experimental study on the design thinking process started from difficult keywords: Modeling the thinking process of creative design. *Journal of Engineering Design, VOL., 14*(4), 429–437. doi:10.1080/09544820310001606911

Nahavandi, A. (2015). Fostering Creativity and Innovation. SAGE Publications.

Pan, L. S. (2009). Diversity of Design Education in China. *Iridescent-Icograda Journal of Design Research, 1*(1), 130–136.

Rubera, G., & Kirca, A. (2012). Firm innovativeness and its performance outcomes: A meta-analytic review and theoretical integration. *Journal of Marketing, 76*(3), 130–147. doi:10.1509/jm.10.0494

Staats, L. (2011). The Cultivation of Creativity in the Chinese Culture—Past, Present, and Future. *Journal of Strategic Leadership., 3*(1), 45–53.

Standard Ranking Research Institute. (2015). *Undergraduate Salary Ranking.* Retrieved from http://edu.163.com/15/0428/14/AO9UO15E00294MP6.html

The Free Dictionary. (2016). Retrieved from http://www.thefreedictionary.com/design

Vigleik, N. (n.d.). *An introduction to form-giving in Chinese product design.* Academic Press.

Wang, X. J. (2009). Contemporary Forces Driving the Development of Design. *Iridescent-Icograda Journal of Design Research, 1*(1), 138–146.

Wei, X. H., & Li, Q. Y. (2013). The Confucian Value of Harmony and its Influence on Chinese Social Interaction. *Cross-Cultural Communication, 9*(1), 60–66.

Wong, Y. L., & Siu, K. W. M. (2012). Is there creativity in design? From a perspective of school design and technology in Hong Kong. *Asia Pacific Education Review, 13*(3), 465–474. doi:10.1007/s12564-012-9208-y

Wu, W. P. (2010). *Higher Education Innovation in China.* Richmond, VA: Virginia Commonwealth University.

Yang, W. Y. (1993). Confucius. UNESCO.

Zhang, L., & Dou, J. H. (2011). *Reanalysis of Industrial Design Education in China.* Retrieved from http://dolcn.com/data/cns_1/article_31/paper_311/pind_3111/2011-06/1308229464.html

KEY TERMS AND DEFINITIONS

China's Industrialization: Industrialization means a society or country's transformation from one phase to another, for example, from agricultural society to a manufacturing one that is based on goods and services. Individual manual labor is often replaced by mechanized mass production and craftsmen are replaced by assembly lines. Every country has its own speed on the industrialization process, China is still considered in its industrial phase.

Intellectual Property: Product, work or invention that is usually intangible. It refers to creations of the mind, such as inventions; literary and artistic works; designs; and symbols, names and images. It usually be can copyrighted, patented, or Trademarked through special authoritative organizations and be protected if there is a well-established legal system.

National Art Exam: It is also known as Yikao (艺考). Any students who want to study art, design, acting/drama, theater, and movie field has to take such special exam before National Entrance Exam. It usually occurs in late Jan or early Feb after winter break or Chinese New Year. Different schools would hold their own test during a specific time at a regulated location in different provinces. All exam results will be collected by each location and send to the specific schools for grading and selection.

National Entrance Exam: It is also known as Gaokao (高考, "Higher Education Exam"). It is a two-day academic examination held annually on June 7-8th in China. This examination is a prerequisite for entrance into almost all higher education institutions at the undergraduate level. It is usually taken by students in their last year of senior high school. Every student (including art students) has to take the exam and receive a score to be evaluated during admissions office by the universities they applied.

Preparatory Training Studios: Known as Kaoqianban(考前班, "Pre-College Studio"). It is usually privately held studios by graduated students from top art and design academies. They train senior high-school art students to prepare for the art and design exams for different universities. Studio usually starts in the summer (July-Sept) and closes after students completes all art exams the next March.

Social Design: Design that intends to solve a social issue or evoke a good purpose that can serve society well. It suggest a better and more sustainable way to contribute for improving human well-being through a design processes. It can be any form of media thus not limited on what products, form or presentation to be delivered.

Status Quo: The current situation, how things look like now at the moment. It means acceptance of present status or relationship, usually has a negative meaning.

Chapter 2
Critical Thinking and Design Education in China:
Considerations from a Western Perspective

Stefano Ceppi
Stefano Ceppi Industrial Design Studio, Italy

ABSTRACT

"If a science of societies exists, one must certainly not expect it to consist of a mere paraphrase of traditional prejudices. It should rather cause us to see things in a different way from the ordinary man, for the purpose of any science is to make discoveries, and all such discoveries more or less upset accepted opinions" (Durkheim, 1982). The aim of this paper is to describe the surrounding aspects and difficulties for western teachers to teach the important skill of "critical thinking" in China, and to provide a general point of reference about how to do it. Furthermore, this paper's aim is to introduce the development of significant cultural changes in Chinese society in the last twenty-five years. This paper offers a theory of how these changes effect society Critical thinking learning capabilities are deeply related to the community identity and have a relevant influence on many social aspects such as: relationships, behaviour, communication and business. It is also crucial to foster them in design education to promote innovation and the development of emerging nations.

DOI: 10.4018/978-1-5225-0911-0.ch002

INTRODUCTION

...if a science of societies exists, one must certainly not expect it to consist of a mere paraphrase of traditional prejudices. It should rather cause us to see things in a different way from the ordinary man, for the purpose of any science is to make discoveries, and all such discoveries more or less upset accepted opinions. (Durkheim,-1982).

Over the course of two decades or so, China has become the world's largest manufacturer for almost all products used. Design has become significantly important to the economies of many countries, including China, and it is no longer perceived as a cultural phenomenon prerogative of Western Countries. The Chinese Government after many years of emphasis on manufacturing has decided to move to a 'higher' goal; they had started to invest in design education and promoting a culture to encourage a new generation of Chinese designers. Western (dichotomy) education has always been perceived as valuable, due to the technological advancement of Western Countries. With the economic reform in China during the eighties and nineties, many western designers and design-related experts were invited to China to transfer their design knowledge and experience.

The author was fortunate enough to be invited to China as a Design *lǎoshī*, which means "teacher" in Chinese, or literally an "old person of skill". I was a teacher in Raffles College of Design and Commerce in Shanghai, a collaboration between the Raffles Design Institute and the Dong Hua University. The college was a pioneer in the educational field, since it is one of the earliest approved Sino-Foreign joint venture schools in Shanghai. This position gave the author the opportunity to experience the beautiful Chinese culture, to understand the thoughts, aspirations and fears of a generation governed by the single-child policy. For western teachers and experts brought to China, it is fundamental to understand the cultural differences between the Western and the Far East society to be successful educators.

According to Guy, the definition of culture is "shared values, attitudes, beliefs, behaviors and language used within a social group" (Guy 1999, p. 7). The first logical assumption in this chapter identifies *de facto* that western educators and Chinese students do not share the same background and cultural values.

This could lead to misunderstandings due to different interpretations and attitudes, especially in the art disciplines. In addition, the rigid scholar system in China and the policy of "compartmentalization" contradicts the intention of the Chinese establishment to implement a western learning approach. The formation and implementation of the western learning approach, compared to the traditional educational system adopted in China, generates important issues in many scholarly situations especially in the areas that require independent thinking. Critical and creative thinking skills have been neglected in China as such subjects would require a more open cultural system to be effectively implemented.

BACKGROUND

The perception that a better education can determine an individual's success in life is evident in China. Education prepares students with the expertise to intellectually and socially fit into a system to achieve a better personal situation, "Education is generally seen as the foundation of society which brings economic wealth, social prosperity and political stability" (Varun, 2014). In order to obtain such stability, the individual has been "shaped" and influenced to fit into this environment. His behavior, perceptions and assumptions, leads the individual into a socially acceptable manner of thinking (Marquardt, 1999). Clearly in the past decades the difficulties relating to the enforcement of a foreign educational system into China has been underrated. A skill such as critical thinking was not expected to have any relation to the cultural heritage if taught in a classroom. On the contrary, the ability to think critically maintains and reinforces a person's cultural identity. The scholar system in China is inspired by Confucianism, which is not a religion as it does not have the metaphysics introduced later with the neo-Confucianism by Zhu Xi and Wang Yangming, but it still has in his philosophy an intrinsic respect for authority and governance decisions (Lee, 2000),. To the Confucian, harmony present in society and State corresponds to a universal whole, this assumption dovetails with the vision of the Communist Party. As Ellen Avrill (2009) illustrates as Cornell University Asian Art Curator, "during the Great Proletariat Revolution… [Art] … Forced to serve political and propaganda purposes under enormous pressure from guidelines..." (Avrill, 2009). Those cultural guidelines and approach to artistic educational subjects suggest a sense of rigidity and control, therefore an implicit furtherance of a memory learning approach, even for creative subjects.

Caine and Caine (1991, Vii, p. 8) note that a physiological model of memory calls into question the notion that learning must take place through rote memorization. By understanding properties of our spatial memory system, educators can understand that teaching to behavioral objectives ignores other functions of the brain and other aspects of memory and learning. Indeed, we have come to the conclusion that educators, by being too specific about facts to be remembered and outcomes to be produced, may prohibit students' genuine understanding and transfer of learning.

CRITICAL THINKING

It is clear that critical thinking is an interdisciplinary and multifaceted subject, but it is also clear that human creativity has been and will –increasingly-- continue to be, a key factor in shaping our society (Freedman, 2007). In a competitive society where product designs are not only getting more complex but also require of higher levels of technology and skills, mere engineering knowledge is not enough for the evolution of products designed for the human being. This global competition and methodical refinement of industrial process drove the research to going "to stress the academic subjects more" (Lewis 2008). Having gone through economic and social transformations over the last twenty years, Chinese establishments have understood, just like their equivalent western ones did, it is possible to find 'cheap labour' even for high technological or engineering jobs, at least to a certain degree, but it is more difficult to find talent for creative tasks.

It is obvious that teaching creative thinking is vital to better competitiveness and it is without a doubt that, assessing such a skill is not an easy task, it is not like assessing mathematics, where the problem solution is inside the text. Moreover, the academic design world is under much political pressure to address the industry's starvation for new ideas to stay competitive. Creative designing is vital to overcome challenges created by the discovering of new materials with novel properties that could lead to new applications. Therefore, educational research and investments on creativity enhancements are essential. Creative thinking ability is highly regarded as a valuable trait in contemporary societies (Lewis, 2008). Before defining the concept of critical thinking we should take into consideration further social and human dynamics.

Linda Elder (2007) noted, that man is by nature a social animal. Therefore, "... we must interact with others to survive as beings in the world. But we aren't just social animals; we are social animals who form complex belief systems. These belief systems often reflect a variety of forms of intellectual blindness, as well as intellectual insights. We develop worldviews that are a mixture of self-serving, group-serving, and rational thought. Our social groups not only provide us with ways and means of surviving; they also impose on us a relatively narrow way of looking at the world." In China the difficulties in undergoing the study of critical thinking are related to the "socio-centric thought" (Elden 2007), where the vision of the world and any idea is a condition perceived by "the group". Where we could define the group as a "social system" or "society", in general.

Critical approach to design is influenced by culture as well as the experiences and capabilities of an individual. The Chinese government is very dominant in their peoples' life. The first thing that leaps out at a foreigner in China is that the government likes to be perceived as an "over-structured system," one that takes care and guides citizens safely for their entire life. Therefore the Chinese student may not be motivated to speak up or criticize: mere rote-learning without great critical participation leads to a better fit in such a structured society. But the role and influence that China wants to obtain in a global society and the foreign influence over many young Chinese lead to a reconsideration of such an attitude. In the learning and understanding of how to teach a critical approach to design, many considerations must be made, starting from a discussion on the origin of critical thinking.

Definition

The origin of critical thinking can be associated and placed in the Socratic method, well described in the *Theaetetus* by Plato. This consist of a dialogue between the teacher and the student, where the student was challenged to argue and support his own point of view, and find fallacy points in it. In this way the student builds her/his own theory and learns the necessity to always verify personal suppositions. In Eastern countries we can find embryonic traces of critical thinking in Buddha's (Kalama Sutta) teaching, Soma Thera (2013) illustrates this when saying: "The instruction of the Kalamas (Kalama Sutta) is justly famous for its encouragement of free inquiry; the spirit of the Sutta signifies a teaching that is exempt from fanaticism, bigotry, dogmatism, and intolerance." In Europe and in the United States, critical thinking was developed as part of the social and education system.

The evolution of critical thinking passes through Aristotle and his skeptical approach until this day. The Renaissance (15th and 16th Centuries) was a critical thinking epoch, where many students started thinking critically about religion, art, society, human nature, law, and free will. They proceeded with the assumption that most of the domains of human life were in need of searching analysis and critique. Among these scholars were Colet, Erasmus, and Moore in England. They followed up on the insight of the ancients (Richard Paul, Linda Elder, and Ted Bartell, 1997). It is indisputable that critical thinking underlies the history and evolution in Europe, even through a process of democratization of all States and Countries.

The definitions of creative thinking are numerous in literature, and are often field-dependent. They varied in many aspects, meanings and even concepts, although they closely share some associations, like: cultural relationship and skills relationship. The complexity as we observed, differs due to the complexity of the field and culturally conditioned theoretical approaches to the research. As Dwight Atkinson argues, "teaching thinking to non native speakers may be fraught with cultural problems", this underlines a cultural issue, as mainly the foundation of critical thinking lies in Western countries (Atkinson, 1997, Quarterly 31: issue 1, 71–94.). Therefore the teaching of such undefined subject is culturally difficult, even due to the differences already analyzed above.

The Chinese system is more about a "collective society", colliding with the western individualistic approach, which consequently applies a better predisposition towards criticism and debate. Also Atkinson asserted (1997) that, Critical thinking may be more on the order of a non-overt social practice than a well-defined and teachable pedagogical set of behaviours; (Atkinson, 1997, Quarterly 31, issue 1: p. 71-94). If critical thinking can be considered a "social practice," it already highlights the fundamental differences in terms of societal views between the Eastern and Western Countries. This is especially so in China, where the "communalistic" approach allows the maintenance of a stable and peaceful society.

Ennis, R., & Weir, E. (1985) defined critical thinking as a "reasonable and reflective thinking focused on deciding what to believe or do". In other words, critical thinking can be considered as a "sceptical" attitude towards what we trust or may not trust while performing an action. It is a procedural disposition. When reading a piece of news in the paper, in a Western Country for example (as critical thinking is influenced by culture and social beliefs), such news can be subject to criticism, and readers form an opinion, debating or listening to others with different viewpoints.

This is naturally difficult in countries where there is a central control of information and a univocal source of facts and events. This does not mean that Chinese students cannot become critical thinkers, but they would probably need a period of adaptation before they can integrate into a western education system, where "rote learning" (Bloom 1956) is not the preferable path available. On the other hand, professors working in emerging nations are required to supply a flexible design teaching methodology. Already in 1950, Joy Paul Guilford asserted, in his articles for the American Psychological Association (Guilford, 1950, p. 446), that both industry and governmental agencies are also looking for leaders; men of good judgment, planning ability, and inspiring vision are in great demand. In recent years due to the economic and social changes, China is faced with this demand.

D'Angelo (1971) considered rightly that critical thinking is a process of evaluation. This process consequently involves different social areas and skills. This allows an industrial designer for example, to come to a logical solution to a problem, or to a "neutral" judgment of it, which implies multifaceted aspects and a complex mental evaluation process. The cognitive domain involves knowledge and the development of intellectual skills (Bloom, 1956). This requires the interaction and mastery of many skills, such as: comprehension, analysis, evaluation, synthesis and evidently, knowledge. The promotion of a higher level of thinking is necessary in education and in society, especially in the digital era, where --in most countries-- access to information is almost unlimited. In China, this evolution and its relevant implications are a concern for the government. The intellectual and sociological problems lie between the respect of the tradition (strongly influenced by politics) and the evolution of a cultural system, also called "progress". As summary of many sometimes overlapping definitions, we can describe critical thinking as a set of skills to control and process information easily and efficiently in order not only to perform well in a critical education system, but also in life.

INNOVATION DYNAMICS

In China many companies had started to implement the industrial ISO standards throughout the SAC organization (Standardization Administration of China), established by the Chinese Government in April 2001. This is so that China can be competitive with western companies in the production of refined and reliable products. Another system used by Chinese Companies encouraged by the Chinese government is the acquisition of European or American Companies to develop rapidly. These strategies provide an enormous reservoir of resources and innovative talents.

This new approach requires innovation in terms of design, quality, reliability, and conformance to international standards, environmental considerations, cultural implications and product life-cycle. The selection of a creative worker capable of making such decisions could affect the quality of a product or of a Company in a critical way. Creative thinking is implemented and applied to "HOW" components could be reused, designed and experienced. Critical thinking then is strictly connected to the development of new products and to lead in design development. From the success of design-oriented companies like Ikea or Apple, it is clear that the extremely high importance place on design and of the capabilities to differentiate themselves trough a critical approach to design, is a key element to set the standards and become leaders in the global market. The individual can play a role and have a tremendous impact on the design and development of a firm. Steve Jobs is an example of how a single person with a critical thinking attitude can drive a company to be a Global leader and set the pace in trend and innovation. This of course is characterized by a strong individualistic focus present in western societies.

On the one hand, the issue of unlimited access to information by Chinese students and people could lead to a "disharmony" inside the PRC, but on the other hand, it is necessary to let students, researchers and people acquire the skills to select, analyze, synthesize and evaluate enormous numbers of data and information, since China legitimately aims to sustain its economic growth and technological leadership. The aim of 2006 MPL (Medium-to Long-Term Plan for the Development of Science and Technology), set by the Chinese government, was to reduce China's reliance on imported technology to no more than 30% within a few years. This includes increasing domestic R&D funding, and to leapfrog foreign rivals in what the government identifies as "strategic emerging sectors," (Abrami at al, 2014, Hardvard Business Review). The problem is how to promote individual knowledge in a collective social system, without losing the *status quo*.

According to the author's experience; students' success could consequently enhance the Country as we have experienced, and also the university's reputation and fame. Furthermore, the high level of professionalism and standard reached by the university would lead to a better development of their R&D Department and the registration of new patents. These innovative and unpublished technical information are becoming increasingly important avenues for securing industry financial support for the universities. This information is vital for emerging nations, as it directs industrial decisions and business strategies, since it cannot be fully achieved in a memory-based scholar system.

Graduate students in many western universities are expected to work independently, particularly in their final year, while still within the 'protected' college environment. They are challenged to pick a topic, work with it and present their assignment. They have to critically discuss their work in order to arrive at an improved product or a new perception of the topic which in turn demonstrates if they have achieved the maturity of a graduate student.

Such maturity is recognized as readiness for the student to enter the work environment and society as an adult. In China this maturity is *under construction*, in the attempt to compromise between the Confucian influence on the traditional Chinese education system and a modern society. As Guo and Blanchard (2008) illustrate in *Harmonious World and China's New Foreign Policy;* China is "a country that has gained tremendous confidence due to its political and economic successes[1]" but it is also clear that the cultural "evolution" is difficult to go along with the cultural "revolution".

Awareness

What this author noticed was that some students like to fit comfortably into the system, either because they are already under heavy pressure from the family's expectations, or because they are simply spoilt as a result of receiving the entire family's care and attention.

Critical thinking in the Western approach does not imply cultural superiority or intellectual knowledge in the general sense. It is simply a way to develop rational rules regarding problem solving through objective analysis that is coherent with the cultural environment. Due to the tremendous determination of the Chinese people and global exposure of the younger generation, changes are imminent. Still an important debate for China is to find a compromise between the modern society they would like to present globally, and the tradition ingrained in history. It was well stated by Liu Yongxing (Hope Groups) to the famous American business magazine Forbes: "Ironically, the survival of the Chinese communist party depends on a healthy entrepreneurial class. Within the next ten years, private businesses, including foreign companies, may become the most important components in China's economy." So the vital importance in the use of critical approach to thinking, fundamental to the Country's economy, must be supported and better understood by the academic system. Those young Chinese exposed to foreign stimulus will be part of the élite that will guide and renovate the Country in the next decade.

There are a number of students that possess natural critical thinking skills. At least they are aware of it, but due to the enormous difficulties to express themselves in English, they find it more convenient not to challenge themselves. The problem could lie with the 'methodology' (Guilford, 1950), since in China the grade is the most important concern in a student life: it determinates the chance to progress socially, which will be analyzed further in this chapter. In art subjects, the potential of creative thinking could lead to outstanding outcome, as those artistic fields are not constricted by scientific standards. Another important reason for the neglect of Course, is the difficulty of the problems themselves. A practical criterion of creativity is difficult to establish because creative act of an unquestioned order of excellence are extremely rare (Guilford, 1950). Guilford (1950) stresses the point about the so called "accidental nature of many discoveries" could drive us to question how rigid an education system should or could be.

Compartments

From the author's experience in China, many students were focused on their grades, without really understanding the subject matter. This, as R. Abrami, W. Kirsby and W. McFarlan mention, is partially the result of what Japanese scholar Ichisada Miyazaki calls "China's examination hell." How can students so completely focused on test scores possibly be innovators?" (as cited in Harvard Business Review, 2014, p. 1).

While the aim of critical thinking is learning "through experience", in China the concept of "compartment system", implemented and used for many decades, resulted in an education system that is very much focused on assessments. Teachers and students prepare themselves for school tests, and "knowledge" and information not assessed in the test are not considered important (Zhen, 2006).

The paradox is that in China's ancient culture, we can find many examples of critical thinking; let's simply think about the philosophy applied to traditional martial arts, for instance, the aim of acquiring martial skills is the experience itself. The animal movements used in Kung-fu (wushu - 武术) must be selected, studied, analyzed, decoded, imitated and implemented in a creative way before obtaining the "forms" necessary to create an effective and innovative decoded defence movement or attack. This well matches Bloom's taxonomy of learning objectives (1956). Since critical thinking is originally a cognitive activity related to the mind and not to instinctive reactions to events. Learning Kung-fu is an example to explain that certain forms of critical thinking are present in Chinese culture and DNA. But with the "compartments system", somehow, all different skills are kept in boxes which are not always applied or connected to other skills and "boxes" to obtain a multifaceted autonomy.

As discussed above, the Chinese education system has for long been well controlled and structured to conform with the Communist Party philosophy. But this fact contradicts with the freedom required to conceive ideas in arts. In fact, since the time of the Cultural Revolution, art has been perceived as an instrument for propaganda, and is somehow considered dangerous. This ancient prejudice towards art disciplines doesn't affect the implicit application of critical thinking to other disciplines like, for instance, martial arts, which are probably less perilous for the stability and "harmony" of China (Lee, 2000).

To be fair, it is important to consider also the extraordinary dimension of China's territory. In order to govern and organize the vast population in China, a structured and univocal education system was deemed to be necessary. The communist philosophy also undermined the "democratic" possibilities to reach success for everyone, through a fair scholastic path, guaranteed by the Government regardless of a person's background. However, such highly regarded communist philosophy required a standardization of a marking system throughout the Country, and therefore even a codification of subjects taught in colleges.

The question of lateral thinking has already been debated within the western education system. Guilford states that: "It should be remembered that from the time of Binet to the present, the chief practical criterion used in the validation of tests of intellect has been an achievement in school. For children, this has meant largely achievement in reading and arithmetic. This fact has generally determined the nature of our intelligence test" (p. 447). Nowadays critical thinking applied to design is developing even further, from a mere "aesthetic sense" (Munari, 1971), to an experienced state. Customization already requires an in-depth understanding of knowledge and application of critical thinking. Design, which is a generic umbrella term under which many specific disciplines operate, (graphic, vehicle, product, multimedia etc.), is shifting to UX (User Experience Design), which is mainly concerned with "how the product is felt or used. This already emphasizes the importance of critical thinking as a process. It is evident, for example, in the attempt of grasping and analyzing and evaluating what people feel, want, perceive and do but communicated in a non-verbal way. Feelings are intangible; they are involved in critical thinking and in the design process of new products and new high-tech products. What the author noticed was the lack of critical observation skills and creative innovation attitude demonstrated by Chinese students when evaluating a situation which often just skimmed the surface.

The author's experience teaching critical thinking is not using English but through translators available at a College and who not only had an excellent knowledge of the language but also were experienced in design subjects. This is an important consideration, since language plays a pivotal role and can represent an intrinsic cultural obstacle to a fair learning process.

In the implementation of critical thinking applied to design and design education, the methodological approach plays a significant role. Although critical learning in China and in other emerging nations is different from western countries, from a cultural identity and linguistic prospective it is still feasible to transfer to Chinese students a critical attitude in learning and to build a unique personality through a responsible approach by motivating their interests.

Educational Environment

The teacher plays an important role to "harmonize" the learning of critical thinking in a scholastic environment. To offer young Chinese students a different prospective and to develop a mature use of critical thinking, as Richard Florida illustrates: "the process of destroying one`s gestalt in favor of a better one" (Florida, 2002, p. 31). Where "better one" can be interpreted more appropriately for the purpose of this paper as: the ability to think critically.

The author and other foreign teachers were inducted into their roles as teachers in China, they were instructed about the importance of writing on the white-board whatever materials the students are required to bring to class or they risked the students not being fully equipped for the job they were expected to do. This episode revealed that the Chinese education system is reinforcing a doctrinal system to apparently support students. In reality, there could be an oversight in this type of education system with regards to the needs and requirements to appropriately instill critical thinking. The author noticed that this over-protection was not productive and functional to the process of teaching critical thinking which agrees with the critics that Guilford lavishes to the western education system during the '50s.

The teacher's role changed from one that 'owns the truth' and teaching it to the students, to an educator who guides and can be questioned, someone who will have a true interaction with the students. The role of acting as moderator rather than the judgmental teacher was the first key point to introducing critical thinking to students. The experience and environment must be perceived as friendly and should encourage the students' to voice their views.

Prescribing a list of necessary drawing tools but omitting to write on the board pencils for the next drawing lesson implied that in the following lesson, many students (not all), came without pencils. They came with all the materials written on the white-board: pantones, rulers, eraser, French-carves and pen but not a single pencil. When the students were asked the reason why they did not consider important to bring pencils for a drawing lesson, their scared and timid answer would be that the teacher did not mention or wrote pencils on the 'to bring' list on the board. The "compartmentalization system" and the cultural/education background will take time to integrate with a flexible style of teaching. This example well illustrates the importance of understanding and knowledge of the Chinese cultural background for many foreign teachers to interact appropriately with Chinese students.

Another aggravating issue was that some foreign teachers left China too early. Many were probably frustrated by certain misleading student attitudes, as they were neither prepared to face such a challenge nor did they have the sensibility to consider the context or decoding the different behaviors. Beside the indisputable success of Chinese students in the Program for International Student Assessment tests, the Chinese Government is also perfectly aware of the need of making the local educational system more autonomous. In fact a recent reform has been discussed and has led to a revolution of the Gaokao exam, introducing more flexibility and giving to students many more choices. The new reform allows students to choose more freely, majors. The new flexibility interested also the university admission, were a "matchmaking" will replaced the score-based tiers exam, as Professor Yong Zhao from Oregon University well illustrated, talking about the positive effect of the reform on the system by: "de-emphasizing the value of test scores and granting more autonomy to students and educational institutions" (as cited in Shore light Education blog by Danxi Shen). Professor Yong Zhao also mention in a New York Times interview that: "…Because a narrow education experience that is centrally dictated, uniformly programmed and constantly monitored by standardized tests is unlikely to value individual talents, respect students' interest and passion, cultivate creativity or entrepreneurial thinking, or foster the development of non-cognitive capacities. But it is the diversity of talents, passion-driven creativity and entrepreneurship, and social-emotional well-being of individuals that are needed for the future economy."

The first thing is to identify a strategy to let students understand that critical thinking is a "process of evaluating" (D'Angelo, 1971) that involves, even in class, specific attitudes and experiences to try to teach critical thinking in the most effective ways. The final aim is to develop in students the critical skills to analyze a problem in-depth and not only skim the surface.

Strategy, Practice and Ambiguity

Below is a potential framework based on the previous discussion:

- The teacher should understand the student's educational background, before applying themselves to educating. This predisposition is essential so as not to misunderstand facts and actions and to dissipate frustrations on both sides. The Chinese educational system is based on memory more than criticism (for example: Chinese characters number in the tens of thousands), therefore it can be difficult to undertake such a paradigm shift. In contrast with critical thinking subjects, it is important for the western teacher to make Chinese students comprehend that examination is also based on the quality of classroom experience, much more then memory tests.
- The teacher should understand the relation between local culture and critical thinking. Chinese students comes from a culture where the holistic approach, for example in the practice of traditional Chinese Medicine (where the whole person is treated vs symptomatic cures in western medication), is widespread and complex.
- In a Country of over one billion people, the competition is high, therefore failure is not an option. A single bad school result can affect a student's entire life. The single-child policy also puts enormous pressure on students. Also many middle-class Chinese families perceive expensive western education as an investment that will benefit the entire family. The foreign teacher should understand and help to alleviate such pressure from many students.
- Fitting into the system is a requirement. A teacher in China is perceived as an "authoritative" figure, and cannot be challenged by students, as that would undermine the authority figure. It is important to build a trust with students by questioning them in an encouraging way. It is not necessary to be judgmental or frustrated about the difficulties in getting the right answers. It is important to encourage a dialogue. The teacher must be perceived as being respectful of other peoples' ideas, even if they have not met the lesson objectives. To stimulate the learning of critical thinking and remove the dread of the teacher being the provider of information, it is important to engage the students' attention through constant interaction and promoting discussions

during presentations. Let students select key information and write personal notes. Do not always tell them what and how they should write notions and information.

- As the environment 'moulds' a person it would be unrealistic for a foreign teacher to expect foreign concepts or methodologies to be understood by Chinese students in just a few lessons and meetings. It takes time to change and break cultural mechanisms which are different (not wrong) from a western understanding of teaching. A foreign teacher would think automatically in a certain way, which is the system that they have always used in their past. This also applies to Chinese students: to effectively teach critical thinking, the teacher must consider NOT only his/her methodology but also the application of it in a different context.
- The feedback will not be immediate as Chinese students will accept the methodology slowly but at the same time, will not complain or question an older person who is in authority. The silence doesn't mean that the students had fully understood or accepted what the teacher had said. A western teacher must understand that the silence or the fact that there are no complaints meant that they had won a scholarly argument. Many students in China may know the answer to your question, or would even like to challenge or critically evaluate a topic or object, but since English is not their main language they are likely to avoid challenging you and risk 'failing' or be embarrassed.
- Critical thinking applied to work requires a vivid and sharp mind-set to resolve problems after having analysed the subject. If the educator gives material to Chinese students to be analysed, it is possible that they will require more time to delve into the question. It is not only due to language barriers as the author experienced that even if the material were in Chinese or explained with a translator, it does not change the amount of time required to look into a challenge. This is due to the Chinese education system that is based on test and the fear of failure or of appearing silly. Brain-storming is a difficult concept to be developed at a professional level in Chinese colleges, even if the teacher's expectation from young students is to find a creative hotbed. It requires time. For a Chinese student to critique a classmate's work is difficult, as they feel unprepared about the subject that they had just been introduced to by the teacher.
- Critical thinking contradicts the notion of a 'compartment system' used in China. It is important to promote to students multiple interests and causes, as creative people and future leaders usually have a broad education in subjects which lie outside the main faculty subjects. Due to her/his cultural back-

ground, the foreign teacher could unintentionally challenge students with themes that they are probably not used to discuss at home. Challenging a Chinese student to promote critical thinking through political or religious beliefs could be counter-productive. The emotional response could be misleading. It is undeniable that emotional feelings can lead to a debate and is a natural approach to critical thinking. In fact, for certain sensitive subject matter, it usually helps to separate "opinions from facts" (Mayfield, 2001), for example, talking about certain topics in China could be perceived as interference.
- Critical thinking is a process linked to problem solving. It includes verification of evidence given or, even better, if defended by the student. Due to the cultural background, this kind of capability can be hard to implement but not impossible. The foreign teacher should "support" students' ideas and promoting students' discoveries even if initially they will give sterile or partially unrelated results.

Modern Chinese students have a tremendous exposure through video games and virtual worlds, therefore it is important, during the lessons or presentations, to choose interesting images that could promote understanding of the complexity of the lectures. It is important to promote controversy to stimulate debates and the mental process of critical thinking. Motivation is a key issue for any student worldwide, but in the eastern educational environment it has a substantial relevance due to their lack of experience in public speaking.

FUTURE RESEARCH DIRECTIONS

The scale of this important topic is therefore extensive, sensitive and requires a development of a targeted, structured and adaptive policy. To have a better understanding and to allow further assessment of what kind of impact and benefit are achieved in the implementation of Critical Thinking in developing China (for instance, the new Gaokao exam admission process), further researches are required. Analyzing the development of the global internet and the strengthening of links between different hemispheres is important to understand the importance of critical thinking in a global net era. A deep review to define a better and workable strategy, functional to the topic, must be evaluated in the following years. As more Chinese parents and children are exposed to overseas education, the importance of critical thinking has risen, compared with memory route learning system. Moreover, many comparisons can be made between the two methods. Meanwhile, China is working on its educational system and admission process in order to keep in China the bright minds, possibly tempted by foreign education, since they could turn into excellent future leaders. In a modern Chinese society creativity and critical thinking must be nurtured, implemented and studied.

CONCLUSION

Critical thinking is more than just an anthropological or social issue, it is highly relevant to the economic growth of a country characterized by a recent developing phenomenon. China's economy is deeply incorporated and related with what happens globally, even in terms of how it influences domestic culture. A socially prepared, critical thinking environment not only enable ideas to be conceived but also produced, improved and made available new ideas for the global market as well as for the local community. Critical thinking can renew society by shaping a new way of originally understanding and innovatively conceiving design in products.

Western teachers must be aware of the above considerations and examples in order to have a better understanding of the Chinese context and the importance of their role. They must strike the right balance in promoting critical thinking as a universally desirable skill to progress in an academic environment or even unconsciously in the broad and rapidly changing society.

To date, this knowledge transfer has only been partially successful in China, due to the various factors discussed, such as the cultural and political aspects in addition to the extensive turnover of foreign teachers. Some suggestions had been made to address the issue of lack of experience by western teachers when introduced to Chinese culture, with the intention of helping educators to understand the context and the relationship between critical thinking and the social environment. The adaptation of a flexible teaching approach can overcome the language and cultural barriers. Foreign teachers should resist from evaluating Chinese students trough western viewpoints, parameters and believes. Applying a rigid design educational system will result in lower levels of productivity from students and a lower inclination to understand a different system, due to the lack of critical thinking. The percentage of skilled young Chinese designers in the world is increasing, not just based on numbers but in quality. To improve such numbers we must pinpoint and define clearly an open education system to develop critical thinking in Chinese students, which differ in the way we have explained above, from a Western approach.

In fact it would be unwise for China to think that it is possible to improve in "home-made" creative talents basing their education only on a partial implementation of a western education system. A partial implementation of critical thinking would deprive it of any substance. The "compartment system" used by the Chinese Government to conform society and the practice of imposing decisions without the possibility of questioning inevitably influenced the cultural adoption of critical thinking. Critical thinking techniques to generate innovative ideas must be explored by students in a classroom, in a protected environment and workspace that enables self-organization, lateral thinking, debating and interactions but these techniques must also be practiced socially. The teachers must dynamically stimulate and promote understanding and innovation through experience, to better prepare Chinese students to become the next generation of creative designers.

REFERENCES

Abrami, M. R., & Kirby, C. W. (2014, March). Why China Can't Innovate. *Harvard Business Review*. Retrieved from: https://hbr.org/2014/03/why-china-cant-innovate

Atkinson, D. (1997). A Critical Approach to Critical Thinking in TESOL. *TESOL Quarterly*, *31*(1), 71–94. doi:10.2307/3587975

Avril, E. (2009). *The art of China's cultural revolution*. Herbert F. Johnson Museum of Art, Cornell University. Retrieved from: http://museum.cornell.edu/exhibitions/art-chinas-cultural-revolution

Bloom, B. S., & David, R. K. (1956). *Taxonomy of educational objectives: The classification of educational goals, by a committee of college and university examiners*. New York: David McKay Publications.

Bol, P. (1992). *'This culture of ours'. Intellectual transitions in T'ang and Sung China*. Stanford, Ca.: Stanford University Press.

Caine, R., & Caine, G. (1991). *Making Connections:Teaching and the human brain*. Alexandria, VA: ASCD.

D. K. (2014, September 14). *Q. and A.: Yong Zhao on Education and Authoritarianism in China*. Retrieved from The New York Times blog.

D'Angelo, E. (1971). *The Teaching of Critical Thinking*. Amsterdam: B. R. Grunner.

Danxi, S. (2015, April 21). *China's new Gaokao reform and its global implication*. Retrieved from Shorelight Education Blog: http://blog.shorelight.com/chinas-new-gaokao-reform-and-its-implication-on-international-admissions

Davidson, B. (1998). Critical thinking faces the challenge of Japan. *Inquiry*, *14*(3), 41–53.

Day, R. (2003). Teaching critical thinking and discussion. *Language Teaching*, *27*(7), 25–27.

Durkheim, E. (1982). The rules of sociological method. The Free Press. doi:10.1007/978-1-349-16939-9

Elder, L. (n.d.). *Using Critical Thinking to Get Beyond Groupthink*. Foundation for Critical Thinking. Retrieved from: http://www.criticalthinking.org/files/The%20Emancipated%20Mind%20FINAL.docx

Elder, L., & Paul, R. (1996). Critical thinking: A stage theory of critical thinking. *Journal of Developmental Education*, *20*(1), 34–35.

Ennis, R., & Weir, E. (1985). *The Ennis-Weir Critical Thinking Essay Test*. Pacific Grove, CA: Critical Thinking Press and Software.

Ferrero, G. P. (2002). *The Cultural dimension of international business*. Pearson Educational Inc.

Freedman, E. B. (2007). Is teaching for social justice undemoctatic? *Harvard Educational Review*, 77.

Guilford, J. P. (1950). Creativity. *The American Psychologist*, *5*(9), 444–454. doi:10.1037/h0063487 PMID:14771441

Guo, S., & Blanchard, J. F. (2008). *Harmonious word – and China's new foreign policy*. Lexington Books.

Guy, T. C. (1999). Culture as context for adult education: The need for culturally relevant adult education. In New directions for adult and continuing education: Providing culturally relevant adult education (p. 7). San Francisco: Jossey-Bass. doi:10.1002/ace.8201

International Organization for Standardization. (n.d.). *China (SAC)*. Retrieved from: http://www.iso.org/iso/about/iso_members/iso_member_body.htm?member_id=1635

Jaswal, V. K., & Neely, L. A. (2006). Adults don't always know best: Preschoolers use past reliability over age when learning new words. *Psychological Science*, *17*(9), 757–758. doi:10.1111/j.1467-9280.2006.01778.x PMID:16984291

Kahneman, D. (2011). *Thinking fast and slow*. N.Y. Farrar, Straus & Ciroux.

Kalama Sutta, The Buddha's Charter of Free Inquiry. (2013). Retrieved by: http://www.chinabuddhismencyclopedia.com/en/index.php/Kalama_Sutta

Lee, T. (2000). *Education in traditional China. A history*. Leiden: E.J. Brill.

Lewis, R. (2009). *Understanding Pupil Behaviour: Classroom management techniques for teachers*. London: Routledge.

Littlewood, W. (2000). Do Asian students really want to listen and obey? *ELT Journal*, *54*(1), 31–36. doi:10.1093/elt/54.1.31

Longman, D. G., Atkinson, R. H., & Breeden, J. A. (1997). *Strategic thinking and reading*. Boston: Heinle & Heinle Publishers.

Marquardt, M. J. (1999). *The global advantage: How world-class organizations improve performance through globalization*. Houston, TX: Gulf Publishing Company.

Mayfield, M. (2001). *Thinking for Yourself: Developing Critical Thinking Skills through Readingand Writing* (5th ed.). Thomas Learning.

Merriam, S. B., Caffarella, R. S., & Baumgartner, L. M. (2007). *Learning in adulthood* (3rd ed.). San Francisco, CA: Jossey- Bass.

Merriam, S. B., Caffarella, R. S., & Baumgartner, L. M. (2007). *Learning in adulthood*. San Francisco, CA: Jossey- Bass.

Munari, B. (1971). *Design as art*. Penguin Books.

Munari, B. (2003). *Bright Minds, Beautiful Ideas: Parallel Thoughts in Different Times*. Consortium Book Sales & Dist.

Nisbett, R. (2003). *The Geography of Thought: How Asians and Westerners Think Differently…and Why*. New York, NY: The Free Press.

Paul, R. (1992, April). Critical thinking: Basic questions and answers. *Think Magazine*.

Paul, R., Elder, L., & Bartell, T. (1997). *California Teacher Preparation for Instruction in Critical Thinking: Research Findings and Policy Recommendations California Commission on Teacher Credentialing, Sacramento California, 1997*. Foundation for Critical Thinking.

Schwartz, S. (1999). A theory of cultural values and some implications for work. *Applied Psychology*, *48*(1), 23–47. doi:10.1111/j.1464-0597.1999.tb00047.x

Svoboda, S. (2015, June 2). *Why do so many Chinese students choose US universities?* Retrieved from BBC News: http://www.bbc.com/news/business-32969291

Svoboda, S. (2015, June 2). *Why do so many Chinese students choose US universities?* Retrieved from BBC News: http://www.bbc.com/news/business-32969291

Tatlow, D. K. (2014, September 14). *Q. and A.: Yong Zhao on Education and Authoritarianism in China*. Retrieved from The New York Times blog: http://sinosphere.blogs.nytimes.com/2014/09/14/q-and-a-yong-zhao-on-education-and-authoritarianism-in-china/?_php=true&_type=blogs&_php=true&_type=blogs&_r=2&

Varun, A. (2014). Article. *International Journal of Education and Psychological Research*, *3*(3), 49.

Žegarac, V. (2007). A cognitive pragmatic perspective on communication and culture. In H. Kotthoff & H. Spencer-Oatey (Eds.), *Handbook of Intercultural Communication* (pp. 31–53). Berlin: Walter de Gruyter.

Zhen, L. (2006). *International Classroom Culture.* Centre for teaching & Learning resources. Retrieved from http: //www.chinabuddhismencyclopedia.com/en/index.php/Kalama_Sutta

KEY TERMS AND DEFINITIONS

Compartment System: An unconscious process of thinking, learnt and assimilated through an educational method hinged on the extreme reduction of a complex ontological argument to a schematic representation based on irrefutable axioms to be slavishly learnt. It is characterised by dearth of capacity to conceptualised and create consequentially an abstract model of a phenomenon, throughout the connection of "bonds" and elements apparently separated.

Compartmentalization: An unconscious psychological defence mechanism used to avoid cognitive dissonance, or the mental discomfort and anxiety caused by a person's having conflicting values, cognitions, emotions, beliefs, etc. within themselves.

Forms: In martial arts the *term form* is used for to identify a set of coded movement used as training in order to learn complex concepts in martial arts.

Kalama Sutta: A discourse from the Buddha contained in the Aṅguttara Nikaya text of the Tipiṭaka, which means and is one of the "three baskets" of teaching that comprise the Pali Tipitaka of the Theravada Buddhism (one of the branches of Buddhism).

"Owns the Truth": Expression that explains teacher perception by students, which is unquestionable about his believes and notion given. All the information given by a Chinese teacher represents the – only – right answer.

Procedural Knowledge: The knowledge exercised in the performance of some task. Procedural knowledge requires a pragmatic disposition.

Renaissance: The activity, spirit, or time of the great revival of art, literature, and learning in Europe beginning in the 14th century and extending to the 17th century, marking the transition from the medieval to the modern world.

APPENDIX

The Institute of Museum and Library Services has collected some interesting statistics indicating that despite the sharp decline in library usage over the last ten years, libraries have actually seen an *increase* in the number of people who come through their doors annually since the beginning of the current recessionary period in 2008. Was this a contradiction, or some kind of error? Table 1 shows the data in question.

Table 1. Visits to public libraries

Year	Per Capita Visit to Public Library	Year	Per Capita Visit to Public Library
1998	4.2	2003	4.6
1999	4.3	2004	4.6
2000	4.3	2005	4.7
2001	4.4	2006	4.8
2002	4.4	2007	5.0

(Data source: National Center for Education Statistics, 1998-2007)[1]

Chapter 3
Developing an Integral Approach for Chinese Design Education: A Spatio-Temporal Framework

Fang Xu
University of New South Wales, Australia

ABSTRACT

Developing Chinese Design Education (CDE) with Chinese characteristics is an old but unending topic that has been discussed in the field of design education in P. R. China for many years. The debate has been constrained by different attitudes towards the nature of design, diverse interpretations of the Chinese cultural tradition and varied perceptions of the role of education. The future development of CDE is eager for new attitude, thinking and approach to rebuild its culture and value system. This article develops a comprehensive framework through redefining the meanings of CDE, applying the principles of spatio-temporal measurement and ADAL model of Integral Theory. This integral approach based on the newly established framework provides a pragmatic and collective method for CDE, not only from a strategic height to simultaneously understand multiple, competing theories and ideas, but also from tactical operational level to holistically manage different perspectives and practical solutions.

DOI: 10.4018/978-1-5225-0911-0.ch003

INTRODUCTION

Accompanying the enormous development in the economic field in mainland China, Chinese design education (CDE) has made great progress over the past few decades, in response to the increasing demand for more designers from the quickly expanding design and manufacturing industries. Since 1998, under the state's education reforms, the guiding ideology of art and design education has shifted from elite education to mass training; from comprehensive art instruction to more job-oriented preparation (Chen & Ma 2011). It led the great leap forward of CDE in the late 1990s. Today China, with approximately two million students studying design or design related courses (Vision Union, 2015), has almost two thousand institutes and universities that can offer art and design programs (China Education Daily, 2016). Many of them are equipped with updated hardware such as new buildings or campuses, large size studios, advanced computer labs and full-scale workshops. Every year, approximately one hundred thousand students graduate with basic skills and knowledge, which helped to ease the staff shortages in many newly emerging design institutions and enterprises. Meanwhile, in response to changes in the practices of design and manufacturing industries, many universities and colleges explore new ways to improve the quality of their courses by seeking co-operation with overseas institutions, pursuing collaboration with enterprises, reforming course content, as well as changing modes of delivery. These efforts have greatly improved the competency level of CDE in such a short period of time.

However, in a modern sense, CDE as an independent discipline in China's higher education system has a very short history. Its approach shifting from *art, crafts, pattern, decoration, art design* to *design* has experienced a tortuous process, in order to develop CDE with Chinese characteristics (Hang & Cao, 2009). A dilemma that has been limiting a thorough discussion for CDE is the ambiguous perception and misinterpretation of design (Jin, 2007). It comes from the different types of education missions that reflected different attitudes towards the nature of design, diverse interpretations of the Chinese cultural tradition and varied perceptions of the role of education. Hence, the discussion the development of CDE with Chinese characteristics becomes a very challenging task.

The past debates have mainly concentrated on both tactical and strategic aspects: one relates to the operational level, the other connects to the ideological level. On the operational level, it mainly relates to the aspects of the course structures and contents (Zheng & Gao, 2015), teaching staff development, practical ability (Su & Huang, 2012), general management, education mode (Qing, 2008; Zheng, 2009), talent-training method (Li, 2010) and the impact of design (Tong 2010). The criticism targets the problems of the ambiguity of long-term vision; absence of its own characteristics; lack of strategic thinking; falling behind the manufacturing industry (Peng, 2007) and disconnection with real world practice (Li, 2015). On the ideological level, it is a concern for the culture and value system of education on a socio-political level. Two types of opinions reflect two extreme attitudes towards the system. The first one believes that the CDE of the future must find a way out and beyond traditional culture, through a revitalization of the spirit of Confucianism, Taoism and Buddhism (Han, 2010; Tang, 2010). It insists that there is much more to learn from these traditions (Cao, 2005; Ma, 2008). But the second opinion believes that the most effective way to improve the practice of CDE is to learn more from the Western value system (Sun & Shen 2013). It criticizes the problem of CDE in that it has not systematically absorbed the successful experience of the Western design education system, since the Bauhaus (Wang, 2008a). Many past efforts, through changes in course contents (Jiang, 2003), modification of the studio projects (Chu, 2006), improvement in delivery (Wu, 2007), collaboration with enterprises and institutions (Wang, Zhang, Pen, Lu & Jin, 2011), exchange with overseas institutes, and so forth, intended to provide the answers to these obvious issues, but they were unable to fundamentally respond to the deeper causes.

CDE often appear chaotic in that many schools' programs do not have a clear vision on both the practical and ideological levels. They have been struggling to deal with the challenges, which not only come from the different objectives of the education missions, but also from the interaction and contradiction among those varied approaches for their objectives. CDE eagerly needs a comprehensive thinking and mapping strategy - an integral approach that can handle the targeted issues in both a specific and systematic manner. The objective of this article is to generate such a potential framework that can develop an integral approach for CDE, in order to holistically and in a well co-ordinated way manage the different views, various approaches and diverse objectives involved in the practice of CDE. Beginning a re-examination of the complicated nature of CDE and drawing inspiration from Integral Theory, this research has developed a meta-framework with a dynamic measurement of temporal and spatial dimensions, which can accommodate any small or big, single or system issues. With the domains of four quandaries and the interaction of the variables of time and space, it can be used to develop a well-defined yet more flexible scaffold. This newly developed integral approach provides an effective thinking tool and significantly improves the mapping capacity, to systematically manage all relevant challenges that might relate to the specific task of CDE. Eventually, a new culture and value system could be cultivated in response to the needs of the current and future development of CDE.

BACKGROUND

A discussion of the path of modernity for CDE has to review its different stages' education missions through re-examining the schooling policy, course curriculum and pedagogy in the practice. Evolution from *Crafts workshop*, *decorative art* to *art design* reflected three different approaches in the development of CDE (Lu, 1996), they were concerned with varied schooling policies - *inheritance education, alternative education, innovation education* (Yu, 2006). The inheritance education is represented as Arts & Crafts, its pedagogy mainly concerns the inheritance of the past; the alternative education is characterized as Art Design, its pedagogy only passively responds to the needs of the current design and manufacturing industries; innovation education is portrayed as Design, its pedagogy intends to proactively face today for the benefit of the future.

Arts and Crafts (工艺美术 *gong yi mei shu*), represents a general cognition of design from the beginning of the last century to the economic reform of the 1980s. As an ambiguous concept, Crafts represents Chinese traditional *hand-making*; Arts denotes the imported Western concept of *design*. Chen Zhifuo, a leading scholar in the area, believed that Arts and Crafts are not antiquated, but are applied arts closely associated with people's daily lives (Hu, 1996). To distinguish the difference between the products of industry, craftwork and antiques, Chen emphasized that Arts and Crafts should encompass industrial products and craftwork. Further more, he proposed that Arts and Crafts should contain three key elements - "aesthetic, practical and economic". Pang Xunqing, a key founder of the earliest design school in China in the last century, defined the distinction between Arts & Crafts and Design (Pang, 1987, 1). He pointed out that "any type of Arts and Crafts all need design, but design can not cover all the contents of Arts and Crafts (Pang, 1987, 37)". Recognizing the special characteristic of Arts & Crafts, he believed that Arts & Crafts in the Chinese context does closely associate with the unique features of the *Decorative Arts* and hand making. He further suggested, "*Pattern* work is to design the form and decoration of all artifacts". Lei Guiyan, an expert in Chinese traditional *pattern* study, expressed a similar view that Arts & Crafts is synonymous with *pattern*. For him, the literal meaning of *pattern* referred to visual image design (Yang & Lin, 1992). He believed that *pattern* should be a general term for upfront design of Arts & Crafts that obeys the rules of beauty. Wang Chaowen (1963), a leading art critic during the time, believed that *pattern* formation is an important characteristic of Arts & Crafts. He differentiated between two types of Arts & Crafts work - one is for practical purposes, the other is for the ornamental.

From the 1920s to 1980s, the pioneers of CDE had made great efforts to rationalize the concept of Arts & Crafts as Design and developed a form of "decorative art theory" with a Chinese characteristic. Particularly from the late 1950s to1980s, led by the Beijing Central Academy of Arts & Crafts, the approaches through "pattern design" and "decorative art" actually demonstrated the passive state of CDE in response to the complicated social, economic and political environment in the Mainland China. Due to the inadequate production capacity (Li, 2002) and consumption shortage under the planned economic system, the inheritance education of Arts & Crafts remained at the level of Crafts + Art. There was no fundamental break-through in the concept of applied art (Zhuge, 2007), and could not "cultivate the soil" for CDE in its progress toward modern design. Liu (1988) criticized that the decorative approach of Arts & Crafts is an oppositional understanding between content and form, technology and art, applied art and beauty, which contradicts the principle of modern design.

The shift from Arts & crafts to Art Design started from the import of three design foundation courses - Plane Composition, Colour Formation and Three Dimensional Formulation (Tong, 2009), as well as Modern Design History of the World (Wang, 1995) in the early 1980s, after long-term dis-connection from the outside world. The emphasis of the courses was to develop students' comprehensive formation capacity of art creation, judgment and perception (Xin, 1999), which tries to start the approach of Art Design from its new foundation. In the middle of the 1980s, several leading art design schools sent their key teaching staff to developed countries and regions for visits and short-term study. The different teaching philosophies, course contents and methods of modern design education, which reflect the Bauhaus's modernist design principle, were partly introduced to CDE. The approach of Art Design intended to gain the balance of "form" and "function" through the exploration of different styles of form. While it tried to draw a distinction between itself and the decoration of Arts & Crafts, it still wanted to maintain the Art component in design. The establishment of new industry design courses in a few leading art design schools, and the introduction of new subjects, such as Ergonomics, Design Psychology, Materials Technology, and Visual Communication, which greatly extended the scope of the vision of Art Design (Liu 1987) for CDE.

The discussion of "Great Art, Great Academy" in 1995 (Yang, 2016) revealed the desire of many fine art schools to seek future development. The establishment of new design courses in many traditional fine art schools demonstrated a fact that CDE was still seen as art education in China. Even the programs and pedagogies of those newly established design courses in comprehensive universities were more or less similar to those of the fine art schools. Under the slogan of Great Art - "nurture the ideal of aesthetic, not for career education", it could not hide its inherent weakness due to the seriously inadequte contents of technology, materials, processes, and related humanities courses in design programs. The industry asked the design schools to provide students with design skills that can meet the requirements of modern mass production (Gao 2001). In order to help enterprises to reach this objective,

Pan (2003) emphysized that the higher design education should be market-driven and develop service for the professional design talents of industry

The growth of design and manufacturing industries, increase in production capacity and the insatiable desire of consumption provided a new opportunity for CDE. Since 2000, China's higher education entered into the period of *popularity education*. In the following ten years, the government at all levels has been playing a large part in the distribution of government funding to support software and hardware in the construction of art and design schools. A three-stream design education system comprised of comprehensive universities, professional universities and vocational-technical colleges, was established, which can theoretically focus on research-based, application-based and vocational training courses, can offer more than 64 doctoral, 192 master and 214 bachelor level courses or degrees (Chinese Design Education, 2015). The enrollment for the design courses has extended significantly. The schooling policy was dominated by the need to produce useful talents for the emerging design and manufacturing industries. However, the increase in student enrollment in design courses did not ensure the production of qualified graduates (Mathias 2005). Those hastily launched design programs based on the limited teaching resources could not meet the real needs of developed design and manufacturing industries. Meanwhile, the shortage of teaching staff with professional experience restricted course curricula development. A market-driven and job-oriented approach alone displayed unsustainable results (Zheng, Nie, Zhang, & Tang, 2010, 123). CDE needed a more broad perspective on its development. Yuan (2004) suggested two different levels of approach for CDE. In a narrow sense, modern design education is to train students with creative abilities and sustainable development capabilities, students who are capable of designing, teaching, scientific research, production and management. In a broad sense, it is to cultivate human qualities and capabilities that can integrate artistic, scientific and cultural concerns for modern design practice. In doing so, several key art and design schools took the leading role to put the subjects of humanities on the agenda during the adjustment of overall disciplines (Zheng, Nie, Zhang, & Tang, 2010, 207).

In 2010, the GDP of China reached the 2nd position in the world (Nanfang Daily, 2011). However, the overall development strategy of the nation needed to seek more a sustainable and healthy progression. The central government began to promote the construction of "innovative" country, shifting from process and manufacture to design and research; from copy and imitation to innovation and leadership. CDE is finally reaching into the era of Design. In 2011, *Arts* (艺术学*yi shu xue*) was a separate category for the first time, independent of Literature, and it became a thirteenth category along with 12 other categories (China Education Network Channel, 2011) in the new standard classification of disciplines issued by China's Higher Education Committee. Being upgraded to a second-class under the *Arts* category, *Design* as a newly emerging discipline, it has gained a very important position in the national education system. This change is not an issue that merely concerns the fate of the discipline itself, but is a matter that directly relates to the necessity for the cultivation of independent innovation for a greater rejuvenation of the nation (Wang and Yue 2011). The possible future development of CDE was discussed based on four considerations: history and the current status of development; academic characteristics and value; research framework and the education system; assessment and the index system (Art News, 2012; Tang, Wang & Jin, 2012).

In the era of Design, design for the servicing of society is continuously being enhanced, social participation in design is also substantially increasing, which becomes the focus for economic and cultural areas. Externally, with increasing international cooperation and exchanges, the number of students studying and returning from abroad has also increased; thus teaching staff with the internationalized background has improved. Many large international academic exchange activities sponsored by the Chinese design schools were put on hold in China, one after another. Almost every school has carried out various forms of programs for international cooperation (Zheng, Nie, Zhang, & Tang, 2010, 207). China has become a new platform for many exchange activities in the field of design.

Meanwhile, communication, collaboration and even competition among the design schools and programs has clearly increased. The key design schools began to seek new approaches in order to develop the characteristics of their own courses. For example, the School of Design in Guangzhou Academy of Fine Arts has developed a distinctive feature, which is to take root firmly in the Pearl River Delta, and service Guangdong's economic development and social progress. The design education, research and practice are closely integrated (Sina, 2010), through encouraging staff to participate in design practice and bring fresh cases into the classroom, while also encouraging students to join the community of design activities in the form of part-time or self-employment in order to grow knowledge and abilities in practice. The School of Design in Jiangnan University has developed its distinctive features, namely, teaching, research and design education systems through the formation of "localization", the "humane", "ecological" and "digital". It focuses on the combination of art and science, tracking the dynamic international forefront of design disciplines, focusing on cross-cultural teaching research and practice, and establishing interdisciplinary practices by "crossover and fusion" (Jiangnan University, n.d.). The School of Design in Hunan University has been trying to integrate digital and international research with practice (Hunan University, 2005). The overall quality of basic skills training for the students is interwoven with teamwork, environmental awareness, a multicultural outlook and design management for the internet era. The College of Design and Innovation in Tongji University shows its big ambition, which is to establish an "international, innovative, forward-looking and research-oriented world-class college of design with great sense of mission of the era, international perspective and native characteristics" (Tongi University, 2013). Four aspects of common values for the courses are identified: "Innovative and inclusive value orientation; Up-and-coming collaborative innovation style; Capability of contributing to the society; Transparent and aspirant college culture". All these efforts demonstrated the different approaches to explore new concepts, frameworks of schooling policy, course curriculum and pedagogy, in order to suit each other's contexts and conditions for new development.

The challenge for CDE today is that on one hand, the life-style and variety within the social thought of modern, Western, industrial civilization has penetrated all aspects of Chinese society, while the Chinese traditional culture and aesthetic value system, based on the individual handicraft of pre-industralized civilization has lost its dominant position. The latest philosophical reflections on contemporary design concepts present new changes, which the future development of CDE needs to synchronize from functional to aesthetic; from unified to pluralistic standards; from technical rationality to the value of the humanities (Zhou, 2013). How in particular, can those foreign design theories and methods respond to the Chinese real world problems? On the other hand, the influence of tradition on CDE is still profound. It is still believed (Wang, 2000) that the first aspect of the education guide is its traditional orientation, which means that education should be rooted in the essence of traditional national culture. In turn, the building of art design education with Chinese characteristics is an inherent requirement. The question is how to overcome the constraints of the negative impact of traditional culture, such as the habit of traditional thinking that inhibits a breakthrough to critical thinking (Zheng, Nie, Zhang, & Tang, 2010, 47). How to integrate traditional handicraft with the modern aesthetic consciousness and contemporary new technology, to enable innovation?

Those long-term debates and newly emerged questions along with their various interpretations are the consequence of cultural exchange and conflict between the "Easter and Wester", "Traditional and Modern" (Chen, 2007). In response, it needs to establish a research mode based on an overall historical view (Wang, 2008b), which could provide a strategic approach to "correctly handling the contradictions between the overall and partial needs on the spatial dimensions; Handling social contradictions between the current and future needs in the time dimension; and handling social contradictions between the society and personal needs in the functional dimension" (Wang, 1998). In order to do so, the critical task is to develop a new framework that can thoroughly examine those fundamental issues embedded in the current debates. To re-examine the unique nature of CDE is an important precondition for its success.

DEVELOPING A NEW FRAMEWORK FOR CDE

What is the Meaning of CDE?

For many people in the fields of design industry and education, the literal meaning of CDE has been taken for granted. But what does it really mean in Chinese? The answer is not as simple as its literal appearance. The validity of its definition depends primarily on its phrase structure, and changing this structure can lead to different meanings. Hence, if the phrase could be structurally dismembered and restructured word by word, it would offer a new insight to discover the new meanings beyond the conventional perceptions.

CDE can actually relate to the definitions of three aspects. The first is about the definition of China, the second refers to the concept of design, and the third suggests the notion of education. While these three terms can have a relatively clear explanation in the English context, they might need to have much more complicated annotation in the Chinese context.

CDE = ? (Chinese + Design + Education)

Based on the hypothesis of this equation, several basic questions can be raised, such as: what is *design*? What is *education*? More specifically, what is the difference between *design* and *Chinese design*? What is the difference between *education* and *Chinese education*? Heskett (2005) used one concise sentence to summarize the complicity of design - " Design is to design a design to produce a design." The same word, *design,* which functions as *subject, predicate* and *object* in the sentence virtually presents four different roles and meanings of *design,* which in turn relates to a general *concept*, an *action*, a *plan* / or *intention*, and the *outcome* / or *result*. It actually extracts the general characteristic of design. Hence, the meaning of CDE may not be as simple as an equation of design plus education.

Secondly, the phrase of CDE can also be deconstructed into separate phrases such as *Chinese design, design education* and *CDE*. These then become pairs of interconnected and counterpart issues, such as *Chinese design* vs. *design education*, or *Chinese design* vs. *CDE*. Therefore an understanding of these three phrases and their relationships needs to be further annotated in order to arrive at the explanation of their denotations.

CDE = ? (Chinese Design + Design Education)

CDE = ? (Chinese Design + CDE)

Further more, *Chinese Design* and *CDE* can both act as a variable or a dependent variable in their relationships. So, the following relationships can be assumed.

CDE = Chinese Design *determines* CDE

CDE = CDE *leads to* Chinese Design

Thirdly, the phrase might become more complicated if it is situated in a specific historical period, using time as a key dimension. To simplify the variation of time, the phrase can be categorized as "CDE of the past", "CDE of today" and "CDE of the future".

If a research project only deals with a single layer of investigation at a *specific point* in the time, then the research question will naturally emerge as (see Table 1).

But when a study engages in the examination of multi-layered issues at a *certain period* of the time, the research questions will become more complicated (see Table 2).

Fourthly, when the relationships of Chinese Design vs. Design Education, Chinese Design vs. CDE, is given the full dimension of the time along with either Chinese design or CDE as variables, the study of this topic can be further configured

Table 1. Issue related to CDE is reasonably direct when it only deals with a single matter at one point

Single Layer Issue	Specific Point	Research Question
CDE	Past	What is the essence of the traditional?
CDE	Now	What is the main issue for CDE today?
CDE	Future	How to develop the CDE that can confidently face the future?

Table 2. The difficulty of study for CDE is increased when multiple two-way issues are involved

Multi-Layer Issues		Certain Period		Research Questions
Chinese design	Design education	Past	Now	How the tradition impact is now? What today should learn from the past?
Chinese design	Design education	Past	Future	How the tradition can contribute to the future? What can the future learn from the past?

Table 3. The complexity of study for CDE becomes extremely challenging, when it involves multi-layered variables at various dynamic points

Perspective	Variable/Dependent Variable		Time	Research Questions
Past	Chinese design	Chinese design education	Past	?
			Now	?
			Future	?
Now	Chinese design	Chinese design education	Past	?
			Now	?
			Future	?
Future	Chinese design	Chinese design education	Past	?
			Now	?
			Future	?

as three types of relationship with nine possibilities. Various research questions might be generated (see Table 3).

Hence, the re-examination of the definition of CDE through disassembling its structure and meaning subverts existing perceptions inherent in the formation of the phrase. The process of such a deconstructive analysis logically derives a series of research perspectives, clarifies the research questions of various layers, and outlines the possible research issues. It virtually creates a much clearer picture of the nature of CDE, and provides a sound foundation for the more effective management of the properties of CDE.

Diachronic and Synchronic Dimensions

Managing the complicated issues involved in CDE, needs a comprehensive model with a well-structured framework and dynamic characteristics, so that the key issues related to the practice of CDE can be clearly defined. In this model their inherent relationship can be thoroughly examined and issues of all aspects effectively managed. The Swiss linguist Saussure (2013) in *Course in General Linguistics*, laid the foundation for the ideas of structure in the study of language. He admitted that the language system should be studied independently, so he identified two different yet complementary viewpoints - *synchrony* and *diachrony* in linguistic analysis. Synchronic method is "... studied as a complete system at a given point in time," which can be represented as an AB axis. The diachronic approach is to track that which is "... studied in its historical development" and which can be represented as a CD axis. He proposed that both should be studied as a complete system at any given point in time, through the combination of the two axes of AB and CD. In another words, the structure that is created by these two axes, virtually represents a specific point of the synchronic and diachronic dimension, so that the study of language can be measured at that particular moment.

Design practice often involves a process similar to the study of language, it needs to seek such a particular moment that synchrony and diachrony may create, so that "a given point in time" can meet "in its historical development". In such a way, diachrony can be used for the study of design *through time* as a dynamic condition. The durational nature of time can virtually represent the evolution of design as a process or a history. Following the development from the past, present and future, it can find some logical relationship between the thousand-year Chinese cultural tradition and design transformation. So the dynamic nature of design related to social, economic and cultural changes, could be fully examined. Meanwhile, synchrony can be used for the study of design at *one point in time* in a relevant stable situation. The broad nature of space can provide a platform for all design outputs around every corner of the world. Crossing various civilizations and ideologies of the world, Chinese design can be simultaneously viewed with the designs of all other nations, when in the same place and no longer isolated. According to Saussure's theory, language itself is a structured system and a self-contained whole and principle of classification. He believed that a good diachronic investigation is always based on good synchronic work. In such a sense, an abstract cross framework can be created, the synchronic dimension can be represented by a longitudinal axis, and the diachronic dimension can be symbolized by a transverse axis. The unique value of this "cross" is that all the relevant issues of CDE can find their places in the framework, and any specific topic about CDE can be redefined and examined with the consideration of relevant variables through the dynamic measurement of synchronic and diachronic dimensions.

Four Quandaries Model of Integral Theory

Wilber (1995) used the cross framework as the base to establish his Integral Theory. In his iconic book *Sex, Ecology, Spirituality*, he attempted to place diverse theories and thinkers in this single framework. It is depicted as a "theory of everything" (Esbjorn-Hargens 2010), which tries "to draw together an already existing number of separate paradigms into an interrelated network of approaches that are mutually enriching." Using a cross as a basic framework, Wilber (2010) presented his Integral Theory as a comprehensive quadrant model - AQAL model. He used five elements to "signify some of the most basic repeating patterns of reality" - *all quadrants, all levels, all lines, all states, and all types*, and attempted to explain how academic disciplines, every form of knowledge and experience fit together coherently in order to be suited to any context and used at any scale. Esbjorn-Hargens (2009), an

Table 4. Wilber's 5 elements and Esbjorn-Hargens's 5 features of the AQAL model

Wilber's 5 Elements (1995)	*All quadrants*	*All levels*	*All lines*	*All states*	*All types*
Esbjorn-hargen's 5 features (2009)	*Map*	*Framework*	*Theory*	*Practice*	*perpsective*

expert in the Integral Theory Program at John F. Kennedy University, believes that the AQAL model is a multi-variant and can be understood in a number of ways. He further summarizes the essential features of each of the five elements of the quadrant model as *map, framework, theory, practice* and *perspective*, offering three different standpoints from which to apply the AQAL model - from a "third-person map of reality, a second-person framework for working within and across disciplines, and a first-person practice for engaging the development of people's own embodied awareness" (see Table 4).

In formulating a more effective method for sustainable design education, DeKay (2011) applies the Integral Theory of Ken Wilber and provides a transformative perspective for integral sustainable design. Although he focuses on only two main aspects of Integral Theory, *perspectives* (quadrants) and *levels* (of developmental structure), DeKay is able to examine the increasing complexity of sustainable design issues through a transcending spiral (Baker, 2012), situating the approach of sustainable design within the context of human developmental levels: traditional, modern, post-modern and integral. Similarly, Fleming (2013) also addresses Integral Theory in *Design Education For a Sustainable Future*, and seeks to uncover the context, values, and behaviors necessary for effective twenty-first century design education. Although the works of both DeKay and Fleming focus on the fields of architecture and interior design, the principle of their integral approach can be interpreted as a general guideline for other areas of design.

The AQAL model provides an integrally informed platform. It enables the researcher to have "more capacity in response to the complexity of real world, and provide a place for including the essential aspects of any given situation" (Esbjorn-Hargens, 2009), and invites viewers to simultaneously understand different perspectives and solutions.

SOLUTION OF AN INTEGRAL APPROACH FOR CDE

To draw inspiration from the essence of the cross framework and the AQAL model of Integral Theory, the integral approach is to develop a holistic and cohesive framework that can fulfill the need to examine the CDE in its special contexts. In doing so, this integral approach can encompass and integrate all the perspectives and disciplines that must be considered for CDE.

The dimensions of synchrony and diachrony create a background with considerations of both duration and broadness, so that all relevant issues can be discussed in *a specific cross point* with the measurement of temporal and spatial dimensions. Meanwhile, using the framework of the AQAL model of Integral Theory is to simply make "free" room in four quadrants, which can accommodate those contributions on their own terms. In so doing, it opens enormous vistas of vision and possibility for discussion and reflective thought. Hence, the integral approach for CDE is a sort of master-key for the integration and implementation of the vast and divergent contributions, and makes it a great companion for all aspiring stakeholders of CDE.

A Cross Framework

The first step for the integral approach of CDE is to create a cross framework by overlapping the longitudinal axis and the transverse axis. Just as the concepts of time and space can be represented through the interaction of synchrony and diachrony, this cross as a framework, can virtually speak for any concept or idea important for the discussion of CDE (see Figure 1). Any existing or new counterpart ideas can be listed here.

Figure 1. The longitudinal axis can present two poles of any corresponding concept associated with the temporal dimension, such as Tradition vs. Future; and the transverse axis can present two poles of any corresponding concept associated with spatial dimension, such as Art vs. Design

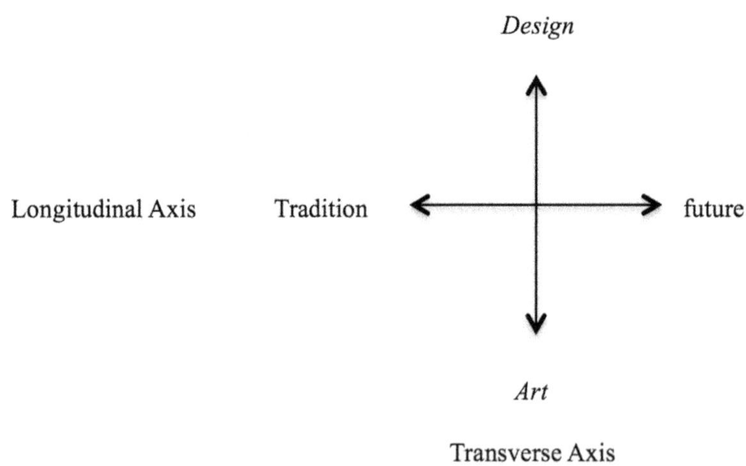

Meanwhile, the field of design and design education in China has been dominated by empirically based perspectives, and limited by a dualistic habit of thinking. The binary-oppositional thought has produced a large number of bipolar concepts, such as *art* versus *science*, *national* vs. *international*, *objective* vs. *subjective*, *past* vs. *future*, *content* vs. *form*, *user* vs. *producer*, *craft* vs. *technology*, just to name a few. These long-standing, controversial issues have arrested habits of thought, to virtually become the main barrier for the development of CDE. The most representative examples are tradition vs. future and art vs. design. Most people cannot "step outside the box" and can only choose between "black and white". However, using the cross framework virtually creates a double bipolar conceptual structure. The previous single, binary-oppositional thought will be replaced by at least dual thought. When the existing single, linear thinking habit is broken, many new thoughts will emerge through comparison and the testing of these multiple pairs of conflicted concepts (see Figure 2).

In addition, the embedded meanings in each term, each bipolar concept, can be investigated further. For example, *Art* is associated with emotions, experience, individual performance, the qualitative, etc. *Design* can be interpreted as rational, empirical, the quantitative, customer-centric, etc. *Tradition* concerns low-tech, practical, hand-made, the collective, etc. *Modern* is related to high-tech, experien-

Figure 2. A large number of bipolar concepts that reflect the binary-opposition thoughts can be selected for parallel and cross testing and discussion

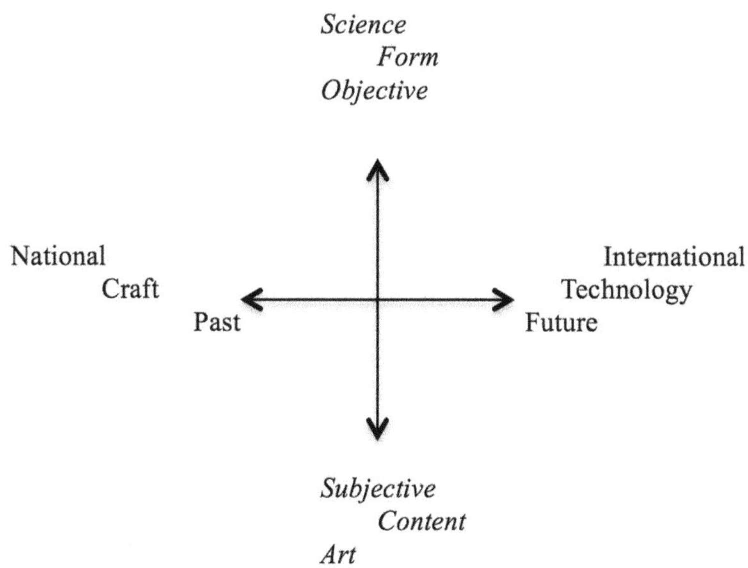

tial, mass-production, individual, etc. The interpretation of these key words can be extended; it can greatly enrich conventional meanings and promote a new understanding. More importantly, it can create new prospects to establish potential relationships with these fresh interpretations. The future approach of CDE would directly benefit from these new opportunities.

Furthermore, between two poles of one axis, it is possible to see a transformative process of dualistic concepts, which can move from left to right, from bottom to top or vice versa. In such a sense, *art* can develop into *design* through the transformation of *craft*, and can *now* carry the value of the *past* of the project into the *future* (see Figure 3). Those concepts of binary-opposition that constrained the cycle of modern design and education can be situated dynamically. With this understanding, the multiple natures of *design* can be further discussed with a better understanding of its relationship to the natures of *art*, *craft* and *decoration*.

Figure 3. The dualistic concepts don't have to compete with each other. To a certain extent, the exploration of the transformative process of the two poles becomes more important than the definition of the concepts of binary-opposition

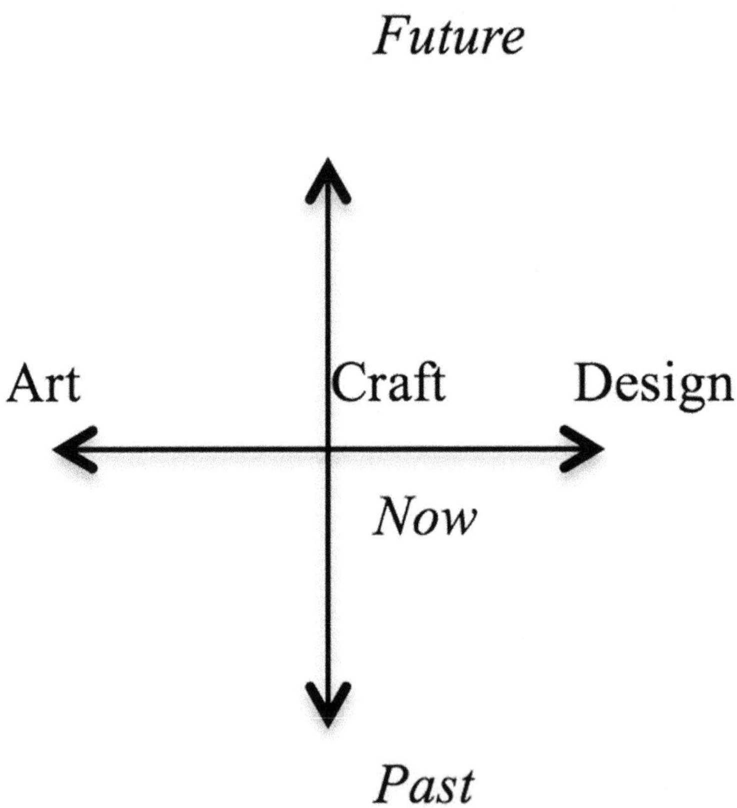

Four Quadrants

The four quadrants that are established by the longitudinal and the transverse axes can be used for CDE as the first move to examine a situation or an issue. Eight different perspectives created by the four domains provide multiple layers of views to process any consideration at hand. Any inquiry that relates to *levels, lines, states, and types of issues* can be thoroughly discussed based on specific world-views, tasks, value systems and the form of artifacts in each quandary. Taking the two pairs of dualistic concepts as an example - *art* vs. *design* and *tradition* vs. *modern* - a framework of four independent domains can be established (see Figure 4). So the issues associated with "five elements" and "features" can be thoroughly examined.

A dilemma that has restrained a thorough discussion of design and CDE in the past is the ambiguous perception and misinterpretation of design among interlocutors due to the different points of views. Hence, this newly proposed framework could offer various domains to accommodate different interpretations, and meet the needs of different interests. Therefore, all annotations related to the multi-meaning of design and CDE can have their rightful places, to be fully examined and discussed.

Figure 4. A proposed framework with selected dualistic concepts can illustrate their specific issues in four domains of LL, LR, UR and UL for CDE

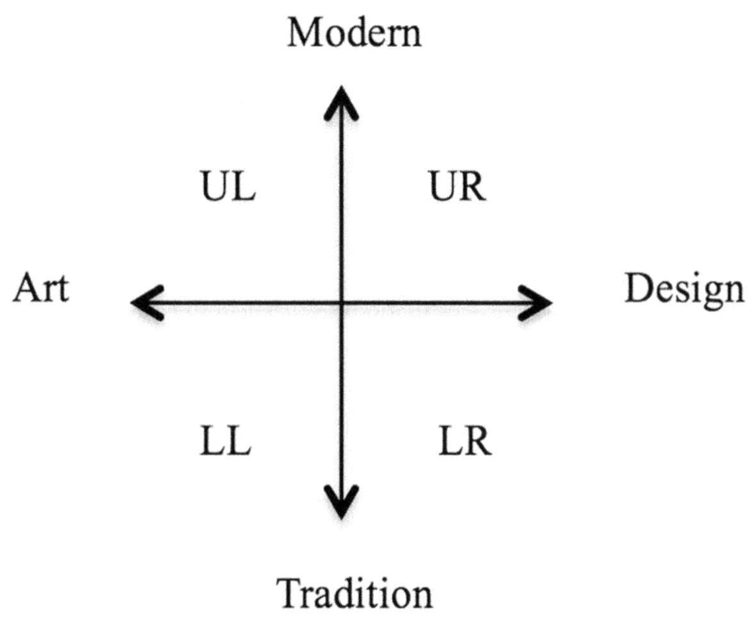

In the lower left (LL) quadrant, the domain can be examined from the dual perspective of *art* and *tradition*. It is the quadrant through which to discuss design issues, which relate to the *craft* interwoven *tradition* and *heritages*, *art* connected to local *culture*, etc. What does art mean to tradition? What does tradition mean to art? What is the essence of traditional handmade craft works? And what is the unique value embedded in traditional artworks? To answer these questions, it is the crucial task of LL domain to seek the characteristics of "design" framed by the tradition and art perspectives, and to explore the meaning and potential of tradition and art as they relate to design. In correlating those tasks, the approach of CDE can exclusively examine the products with their unique tradition and value as art. From the perspective of craftsmen, through engaging in the manufacture and implementation of process and formation, to reflectively learn the logical creations of the tradition, the specific design program for CDE can focus on dealing with material manipulation, processes of fabrication and the development of craftsmanship, in order to draw inspiration from the combination of tradition and art, to protect those rich cultural heritages. For people living in the 21st century, handmade artifacts are more than merely utilitarian, rather they are agents carrying unique memories of the past and emotional connections to their cultural roots. When the handmade can react to design through its low-tech, low-energy approach to consumption, it virtually brings new meaning to modern design practice in today's high-tech era.

In the lower right (LR) quadrant, the domain can be fully examined from the perspectives of *tradition* and *design*: it is the place to examine the issues associated with design practice, as a single professional territory. What is essential for the practice of a design professional? What are the fundamental capacities and skills needed by designers? What can Chinese design learn from the experience of developed countries, since modernism? The domain can specifically focus on an "objective" approach through the development of the techniques and skills of rational thinking and practical making, in order to effectively gain the basic capacity to produce quality design. Although this kind of specialized approach seems out of date compared to the latest cross-disciplinary approach of design practice, it is still meaningful for Chinese design professionals to gradually build their highly specialized capacity. In correspondence to those tasks, on one hand, CDE can concentrate on the learning process through experiencing practical problems, such as the hands-on skills of drawing, modeling and making, so that learners can become familiar with the essentials of form, structure, materials and fabrication. On the other hand, CDE can emphasize critical thinking, to promote the capacity of research, analysis and initiation. As one of the biggest manufacturing nations in the world, so many areas are worthwhile for CDE to develop specific programs, which can make a significant contribution to the industries.

In the upper right (UR) quadrant, the domain can be fully examined from the perspectives of *modern* and *design*. Here is the place to be concerned with the issues about the future development of design; design has to take a proactive approach in response to newly emerging technologies, materials and life styles. After experiencing changes from the creative style, meaning attention, to coordination and management, design development has reached the stage of the creation of the experience. New types of design may be merged, such as that found in a leading approach to innovation, customer-oriented design, solution-oriented design, and so forth. What does technological innovation mean for design professionals? What new issues does creative design have to face? What are the limitations of past design approaches in terms of brand value and commercial success? How can these advanced technologies be of benefit to all people? To answer these questions, it is the main task of UR domain to re-define the new nature of design beyond its conventional perceptions. In correlating those tasks, CDE can exclusively explore the new meaning and new role of design education under the changing conditions of the digital era, adapting the application of robotics, 3D printing, new software and distributed photo-voltaics, and future technological innovations, which provide and will continue to provide tremendous new opportunities for design education. It will be the obligation for CDE to nurture the teaching and learning environments as incubators, which not only initiate cutting-edge design concepts and bridge the gap between China and developed nations, but also push more industries towards a consciously innovative approach. For those design courses in comprehensive universities, CDE could take the advantage by drawing expertise from different streams of faculties to fulfill these complicated design tasks, in a collaborative manner.

In the upper left (UL) quadrant, the domain can be fully examined in two ways, as *art* and *modern*. Here is the place to discuss the issues that relate to the *subjective* dimensions of experience and culture, because any design exercise is rooted in the subjective desire in the human world to change the objective world. When the subjective world is reshaped by new perceptions through men and women of advanced technology, design practices will enter another round of the revolutionary process, from quantitative to qualitative changes. Design can only win people's hearts through poetic creation and excellence in craft. In this domain, the focus needs to be on the often-overlooked areas of the emotional response to design, in order to balance the weight of rational responses to design based on pure high technology. How can design better fit into the local cultural context? How can design transfer symbolic meaning and incorporate cultural connotations? The approach must expand and enlarge the view of design to include the important social and experimental dimensions of human-nature, in order to avoid omitting the realm of culture and human experience. Therefore, CDE needs to fully understand the ultimate goal of design, through the expansion of the connotations and denotations of its practice. It needs to holistically apply the achievements of science and technology, the theories of engineering, aesthetics, psychology and economics to embrace a much needed multidisciplinary and human-scale approach. If the modern approach is to meet the general requirements of the public, then the artistic approach is to meet specific group, social and individual needs: here artistic taste and personal service will have a space in which to expand with a great potential.

Four Ways of Seeing as *Quadrivia*

An integral approach can combine all the insights and outputs from the four domains to become a "four ways of seeing" or a *quadrivia* as described by Esbjorn-Hargens (2009). In doing so, the different perspectives associated with each quadrant are concentrated with a *particular* reality, which is placed in the center of the cross. Once again taking the issue under discussion of Figure 4 as an example, the *quadrivia* can be situated at a strategic height with a macroscopic perspective, or also on a tactical level with a microscopic perspective. Such a kind of setting can be varied, based on the specific goal of the need. The specific value of *quadriva* is that it can also function as a "tuning" device for the main theme of the overall tasks involved in the four domains.

For example, in a macro perspective of *quadrivia*, the center of the diagram of the integral approach deals with the issue of *why*: it needs to stand at the cutting-edge of economic and social development. The theme of *quadrivia* is to set up a strategic vision for future development, such as a shift from dependence on natural resources to reliance on knowledge, innovation and intellectual resources. Positioned in a strategically high place, design should take a leading role in upgrading traditional industries through new technologies, new formats, new models and new industries. With this main theme in mind, the pedagogical philosophy and training mode of design education all need to be reconstructed accordingly.

For another example, in a microscopic perspective of *quadrivia,* the focus of the diagram of the integral approach is to deal with the issue of *what* and *how:* it emphasizes the formative process for a specific design task. The theme of *quadrivia* is to work out a practicable plan in order to fulfill a specific objective, such as the exploration of new techniques for different fabrications, or new materials for artistic value. Standing at such a tactical point, design can concentrate on those issues that only matter to the product itself - new technology, new materials and new fabrication. The multiple natures of design as art, craft and decoration each have their own place to contribute their particular value to the final product.

FUTURE RESEARCH DIRECTION

The future research of CDE can greatly benefit from the dynamic and holistic architecture of the meta- framework. Adaptable to any proposed theme, the critical aspects of existing or newly emerging issues can be unambiguously identified. The possible new approaches of CDE can be thoroughly discussed, predicated on different interests and focuses.

In order to make practical use of it, research priority can be placed on the following three aspects: CDE and China's Design Industry, CDE and China's Manufacturing Industry, and CDE and its three-stream design education system. As discussed in Table 3, many potential research topics related to these three areas can be specified, while the concepts, contents and contexts of their research variables are carefully and clearly defined. For example, as the biggest manufacturing nation in the world, how can the specific programs of CDE promote China's Manufacturing Industry? For these different industries developed under uneven social, cultural, economical and technological conditions, how can CDE's three-stream design programs make unique contributions to their needs? In another case, in response to an emerging historical turning point to promote social change and develop the new culture and value system, how can CDE provide a new direction for China's design industry? How may it assist design practice to transition from passive response to market needs to proactive exploration of future demand?

As a newly developed method, the integral approach requires more empirical cases to enhance its operability. Further research can also focus on one or two institutions, or research a specific design professional, such as product design. It would be beneficial to deconstruct specific cases in line with the issues found, in order to refine the approach of problem identification, implementation process, and the measurement of its effectiveness.

CONCLUSION

The development of CDE is now at a critical point, shifting from what to teach to how to teach, and changing from how to teach to why teach. As a newly emerged discipline, CDE urgently needs to establish its own new culture and value system, which could not only adopt the valuable Chinese cultural tradition, but also absorb the achievements of the Western industrial and post- industrial civilization.

In order to achieve that, CDE has to deal with the challenges from two aspects. The first one concerns the new understanding of the general nature and specific characteristics of CDE; it will have a direct impact on the tasks of the curriculum design construction, pedagogic development, schools' management reform, quality of staff and resource improvement, teaching, research and practising integration, and so forth. The second aspect concerns the interaction of CDE and China's design and manufacturing industries in the changing social, economic and cultural environments. It will influence the position of CDE as both a creator and a beneficiary in the process of creating a new design and manufacturing cultural system. These two challenges are actually interwoven.

The integral approach for CDE is to develop an alternative way that can effectively respond to the above challenges. Drawing inspiration from an essential of the spatio-temporal dimension and the AQAL model of Integral Theory, the simple form of the cross virtually creates a scaffold, which offers a great potential for the new approach of CDE. As a practical tool, all kinds of different views over time can be extracted from two basic facts – the tactical and strategic approaches. The issues on the operational level can be handled in line with the issues on the ideological level, and vice versa. As a new perspective, the existing concepts of CDE can be greatly extended through an analysis their annotation, and a reinterpretation of their new perceptions. As a new attitude, the longitudinal axis and the transverse axis of the cross can break through the limitations of single linear and binary-oppositional thought, to generate multiple points of view for these bipolar concepts. As new thinking and mapping, the territories created by four quadrants of the cross make it possible to focus on issues which are only relevant to one specified domain. While those complicated issues that are less important to the topic can be simplified, it can still maintain the consolidation of the research outputs. As an effective tool, the overall control of the main theme through the chosen point of the cross becomes manageable. It enables the examination of the selected theme to be carried out under the view of the full scope of perspectives. After all, as a new meta-framework with a clear and long-term vision in mind, all existing and new issues of CDE can be more effectively organized, discussed and measured. Eventually, a potential new culture and value system can be established to meet the needs of the current and future development of CDE. So that CDE can develop its own characteristics and more confidently respond to the new challenges of the changing world.

REFERENCES

Art News. (2012). What will bring after upgrading Arts Category? The meeting of history and current situation of the development of art discipline. *Arts News*. Retrieved April 16, 2016 from http://www.zgnfys.com/a/nfms-1978.shtml

Baker, A. (2012). Integral sustainable design: Transformative perspectives, by Mark Dekay: Book Review. *Journal of Ecological Design*. Retrieved March 15, 2016 from http://www.ecotecture.com/integral-sustainable-design-transformative-perspectives-by-mark-dekay/

Cao, L. (2005). *Chinese decorative arts Traditional and its contemporary culture value*. (PhD thesis). Chinese National Academy of Arts. Retrieved February 20, 2016 from http://cdmd.cnki.com.cn/Article/CDMD-84201-2006060477.htm

Chen, H. & Ma, L. T. (2011). Reflection and prospect of the higher education administration reform. *Journal of National Academy of Education Administration*, (6).

Chen, X. H. (2007). Between craft and design: The modernity course of the 20 century Chinese art design. Chongqing University Publishing House.

China Education Daily. (2016). The first higher education quality national report. *China Education Daily*. Retrieved April 18, 2016 from http://www.moe.edu.cn/jyb_xwfb/s5147/201604/t20160408_237162.html

China Education Network Channel. (2011). Arts upgraded to an independent discipline to become the thirteenth disciplines. *China Education Network Channel*. Retrieved December 15, 2015 from http://www.designedu.cn/showposts.aspx?id=299&columnid=7&channelid=1

Chinese Design Education. (2015). Chinese design colleges and universities and courses. *China Design Education*. Retrieved December 15, 2015, from http://www.designedu.cn/curriculumsearch.aspx?channelid=2&columnid=24&selProf=0&selarea=all&seldeg=0&stext=&page=21

Chu, T. (2006). *Design based on insect: Experimental teaching & learning of city design school, Central Academy of Fine Arts*. Beijing: China Textile Publishing House.

De Saussure, F. (2013). *Course in general linguistics*. London: Bloomsbury Publishing PLC.

DeKay, M. (2011). *Integral sustainable design: transformative perspectives*. Routledge.

Esbjorn-Hargens, S. (2009). An all-inclusive framework for the 21st century: an overview of integral theory. *Integralpost.* Retrieved December 10, 2015 from https://integrallife.com/integral-post/overview-integral-theory

Esbjorn-Hargens, S. (2010). Introduction. In Integral theory in action: Applied, theoretical, and constructive perspectives on the AQAL model. Albany, NY: State University of New York Press.

Fleming, B. (2013). *Design education for a sustainable future.* Routledge/Earthscan.

Gao, F. (2001). The development of China's modern art and design in the 2nd half of 20 century. *Arts Observation*, (6), 54.

Han, Y. Y. (2010). Innovation of contemporary art design and inheritance of Chinese traditional culture. *Journal of Changchun University of Technology,* (5).

Hang, J. & Cao, X.O. (2009). Design: another enlightenment - evolution of design ideas and practice of 30 years reform and opening up. *Literature & Art Studies*, (1).

Heskett, J. (2005). *Design: a very short introduction.* Oxford University Press. doi:10.1093/actrade/9780192854469.001.0001

Hu, G. H. (1996). Research of Chen Zhifuo craft thought. *Journal of Nanjing Academy of Arts,* (3), 75.

Hunan University. (2005). *School of Design, Hunan University, School introduction.* Retrieved April 20, 2016 from http://design.hnu.edu.cn/index.php/about/introduction

Jiang, L. (2003). *The Changes of the chair: Overpassing the concepts.* Beijing: People's Fine Arts Publishing House.

Jiangnan University. (n.d.). *School of Design, Jiangnan University, Baike Baidu.* Retrieved April 20, 2016 from http://www.jiangnan.edu.cn/

Jin, Y. (2007). *Chinese design art theory development research after 1980s.* (PhD thesis). Wuhan University of Technology.

Li, J. (2010). *Talents-training mode research of higher art education.* (PhD thesis). Hunan Normal University. Retrieved February 10, 2016 from http://cdmd.cnki.com.cn/Article/CDMD-10542-2010127656.htm

Li, M. (2015). *35 years of China modern products design since 1978.* (Master thesis). China Art Research Institute. Retrieved February 8, 2016 From http://cdmd.cnki.com.cn/Article/CDMD-84201-1015397830.htm

Li, Y. Z. (2002). Outline of arts and crafts. Shandong Education Publishing House.

Li, Y. Z. (2003). Establishing new type of art design education system, developing the course of Chinese art design education. *2001 Tsinghua International Industry Design Forum - Paper Collection of the National Industry Design Education Seminar.* Beijing: Tsinghua University Publishing House.

Liu, G. Z. (1987). *New form of counterparty culture – industry design.* Beijing: Exhibition Design Society.

Liu, G. Z. (1988). History how to inform future. *Zhuangshi,* (1).

Lu, P. T. (1996). Design and decoration: necessary tension – exploring educational road of Central Academy of Arts & Design. *Zhuangshi,* (5).

Ma, Y. (2008). Visual innovation of contemporary art design based on Chinese traditional culture. *Art Exploration*, (4).

Mathias, G. (2005). On China design education from 1990 - 2005. *Journal of Package & Design,* (5).

Nanfand Daily. (2011). China GDP over Japan became the second largest economy. *Nanfang Daily*. Retrieved April 20, 2016 from http://intl.ce.cn/zgysj/201102/15/t20110215_22215153.shtml

Pan, L. S. (2003). Current situation and strategy of development China's higher art education. *Zhuangshi*, (3), 44.

Pang, X. Q. (1987), On arts & crafts: Research of pattern issues. Beijing: Light Industry Publishing House.

Peng, L. (2007). Reflection of contemporary Chinese design education - The status quo of design education and its existing problems in great manufacture country. *Zhuangshi*, (5).

Qing, Z. (2008). *Status quo analysis and teaching mode exploration of university art design education.* (Master thesis). Hebai Normal University. Retrieved March 10, 2016 from http://cdmd.cnki.com.cn/Article/CDMD-10094-2008125100.htm

Sina. (2010). Characteristics of School of Design, Guangzhou Academy of Fine Arts. In *Treat major issues lightly*. Academy Design Graduate Exam Training Institutions. Retrieved April 16, 2016 from http://blog.sina.com.cn/s/blog_61dfedcd0100j543.html

Su, M. Z., & Huang, K. (2012). Practice ability reform and training for art and design. *Examination (Teaching & Research).* Retrieved April 12, 2016 from http://www.xzbu.com/9/view-3413912.htm

Sun, F., & Shen, Y. (2013). The influence of international modernism on China's hundred-year design. *Idea & Design*, (4). Retrieved March 15, 2016 from http://www.cnki.com.cn/Article/CJFDTotal-CYJY201304007.htm

Tang, C. M. (2010). *Three values of Confucian culture on contemporary environmental art design.* (Master thesis). Northeast Normal University. Retrieved March 10, 2016 from http://cdmd.cnki.com.cn/Article/CDMD-10200-2010179169.htm

Tong, H. M. (2009). Introduction: re-cognition of foundational course. In Zhao & Feng (Eds.), Concept: Understanding and expression. Guangzhou: Linnan Fine Arts Publishing House.

Tong, H. M. (2010). Growth of Chinese design. *21 Century Business Review*, (2). Retrieved March 12, 2016 from http://www.cnki.com.cn/Article/CJFDTOTAL-ESYS201002044.htm

Tongji University. (2013). College of Design and Innovation. In *College manifesto.* Tongji University. Retrieved April 18, 2016 from http://tjdi.tongji.edu.cn/about.do?ID=55&lang=_en

Vision Union. (2015). Origin: Design Education and Innovation in the 'New Normal' Society - 2015 National Design Education Academic Forum & 'Design Subcommittee, Chinese Association of Higher Education' Annual Conference. *Vision Union.* Retrieved April 10, 2016 From http://www.visionunion.com/article.jsp?code=201511030002

Wang, C. W. (1963). Love to see and hear. Authors Publishing House.

Wang, M. Z. (2000). Integration, combination and foreseen: on disciplines alteration of Academy of Arts & Design, Tsinghua University. *Arts Observation*, (6), 5.

Wang, S. Z. (1995). *Modern design history of the world.* Shenzhen: New Century Publishing House.

Wang, S. Z. (2008a). Critical analysis of China design education. *Sina Blog.* Retrieved March 8, 2016 from http://blog.sina.com.cn/s/blog_4bdabb4901007t5o.html

Wang, S. Z. (2008b). Critical analysis of China design education. *Sina Blog.* Retrieved March 8, 2016 from http://blog.sina.com.cn/s/blog_4bdabb4901007t5p.html

Wang, T., Zhang, Y., Pen, J., Lu, Q. Z., & Jin, K. (Eds.). (2011). *Unlimited territory: 2011 experimental teaching and learning for graduate design projects in environmental art by four schools' supervisors*. Beijing: China Architectural Industry Publishing House.

Wang, Y. X. (1998). Principle of the development strategies of higher education. *Higher Education Research,* (1), 20.

Wang, Y. X., & Yue, X. Y. (2011). Upgrading of arts and its significance - Professor Wang Tingxin interviews. *The Art Community,* (2). Retrieved April 15, 2016 from http://www.cnki.com.cn/Article/CJFDTotal-YSXJ201102003.htm

Wilber, K. (1995). *Sex, ecology, spirituality: the spirit of evolution*. Boston, MA: Shambhala.

Wilber, K. (2010). Introduction to integral theory and practice: IOS basic and the AQAL map. *AQAL Glossary, 1*(3). Retrieved December 12, 2015 from http://www.humanemergence.nl/uploads/2011/03/IOS-Basic-Intro-to-Integral.pdf

Wu, X. F. (2007). *Want to do design: Training program to rapidly enhance innovative capacity*. China Social Science Academic Press.

Xin, H. Q. (1999). *The Method of shape & form formulation*. Hangzhou: China Academy of Fine Arts Publishing House.

Yang, C. Y., & Lin, W. L. (1992). Lei Guiyuan on pattern art. Hangzhou: Zhejiang Academy of Fine Arts Publishing House.

Yang, X. (2016). Yang Xiaoyang. *People Weekly,* (10). Retrieved March 16, 2016 from http://www.wtoutiao.com/p/10btIOK.html

Yu, Q. (2006). General theory of design science. Chongqing: Chongqing University Publishing House.

Yuan, X. Y. (2004). Research of development path of China art design. Beijing: Beijing University of Technology Publishing House.

Zheng, S. Y., Nie, Y., Zhang, B. M., & Tang, L. T. (2010). The Development strategy research of China' art design education. Beijing: Tsinghua University Publishing House.

Zheng, X. N. (2009). Establishing innovative education mode for art & design universities. *China Higher Education,* (21), 56.

Zheng, Y., & Gao, X. (2015). Research and practice of innovative talent training mechanism of Comprehensive art colleges and universities. *China University Teaching*, (12), 22.

Zhou, X. (2013). Philosophical reflections on contemporary design concepts. *Zhuangshi*, (6).

Zhuge, K. (2007). Inheritance from the fission. Chongqing: Chongqing University Publishing House.

KEY TERMS AND DEFINITIONS

AQAL Model: It is one of the most comprehensive meta-theories from Ken Wilber's Integral Theory. It is a multi-variant model with the features of map, framework, theory, practice and perspective. According to Esbjorn-Hargens (2009), all human knowledge and experience can be placed in this model; it can explain how academic disciplines and every form of knowledge and experience fit together coherently.

Arts Category (艺术学*yi shu xue*): It was separated from the category of Literature and became the first-class and independent category in the New Standard Classification of Discipline Categories issued by China's Higher Education in 2011. Accordingly, design was then upgraded to a second-class discipline under the Arts category and gained a very important position in the national higher education system.

Arts and Crafts (工艺美术 *gong yi mei shu*): A special phrase that represents the cognation of design from the beginning of the 1920s to the economic reform in the 1980s. As an ambiguous concept, *Crafts* represents traditional Chinese *handmaking*; *Arts* denotes the imported Western concept of *design*. For academics, it is decorative art; for the general public and politicians, it is traditional crafts and related artifacts.

Art Design (艺术设计 *yi shu she ji*): As a government official recognized and statutory word, it refers to the concept of design in a modern sense in Chinese. It presents design as a phrase that combines art and design together, in order to distinguish design in the areas of pure engineering and technology.

Chinese Design Education: A newly emerging discipline in Chinese universities and colleges since 1990s. It covers a wide range of courses and programs related to design and manufacturing industries, including visual design, industry design, spatial design, animation & multi-media design, art & craft, fashion design, product design, environmental and interior design, and so forth. Chinese design education in this article only refers to the situation in Mainland China since 1950s and does not include those in Taiwan, Hong Kong and Macao.

Diachrony and Synchrony: Swiss linguist Ferdinand de Saussure theorized the concepts. In this article, it is presented as a simple cross framework, and used as a structure that can support any relevant issues with the consideration of spatio-temporal dimensions. It provides a base to create a four-quadrant grid for an integral approach.

Integral Approach: It applies the lens of Integral Theory to the complicated issues of Chinese design education; it moves beyond the current thinking model and embraces a much needed multidisciplinary and contextualized approach.

Chapter 4
So, What is Creativity?
Animation Students and Teachers' Conception of Creativity in China

Henry Ma
The Hong Kong Polytechnic University, Hong Kong

ABSTRACT

With the impact of the development of the Creative Industries in the UK and other countries, China adopted the term Cultural and Creative Industry in a national strategy to initiate a structural refinement of the industrial sectors. The animation industry in China has long served as a major original equipment manufacturer (that is, producers of contract work for an external brand) to foreign investors. It then started to develop into a center of original design manufacturers (self-originated work sold to others) and original brand manufacturers (self-originated, self-branded work). This led to a rapid demand for creative talent from higher education institutes. Creativity is a relatively complex concept, and successfully fostering creativity in education demands a clear conception of what creativity and creativity education are. The objective of this chapter is to explore how teachers and students perceive the meaning of the term creativity and identify factors that contribute to teachers and students' conception of creativity in education.

DOI: 10.4018/978-1-5225-0911-0.ch004

INTRODUCTION

Animation education has developed at a fast pace in China over the past ten to fifteen years. In 2000, there were only two higher education institutes providing animation education programs. By 2010, more than 1,000 higher education institutes were providing about 2,000 animation programs in China (Xiao, 2011). The rapid expansion of animation education was due to the intensive support for the development of the Cultural and Creative Industries by the government. The animation industry, as one of the key Cultural and Creative Industries, benefited from such a move (Wang, 2012).

The tremendous support and investment in animation education resulted in the rapid growth of the number of animation education programs and of the student intake. The increase in the number of programs and students seemed to be a success, but on the downside, it also created lots of problems in the animation education system (Bo, Jing & Liu, 2009). In 2013, a search with the keyword "animation education" in the China Journal Full-text Database returned more than 400 articles, and half of them discussed development issues in animation education. Most of them claimed that animation education could not produce enough talent to match the demand from the animation industry, and identified a serious shortage of qualified animation teachers. A large number of articles also touched upon words like creativity and innovation, but no article actually focused on nurturing creativity in animation education in China.

Animation is one of the key sectors of the Cultural and Creative Industries. If creativity is considered a major component in the Cultural and Creative Industries, nurturing creativity should play an essential role in animation education. However, in the vast number of published articles on animation education, we seldom see research topics addressing this issue. This reveals a potential knowledge gap in the nurturing of creativity in animation education. The study of nurturing creativity in animation education will contribute to achieving a greater understanding of the development of animation education as well as the promotion of the Cultural and Creative Industries.

The main focus of this research project was to explore the current situation regarding the teaching objectives and approaches of higher education institutes on the subject of animation education, and try to understand how teachers and students perceive the meaning of creativity in that field.

THE RISE OF CREATIVE INDUSTRIES IN CHINA

The term Creative Industries was a concept that originated from the Tony Blair Labour Government's establishment of the Creative Industries Task Force by the Department of Culture, Media and Sports (DCMS) in the United Kingdom in June 1997. The Task Force designated 13 key sectors of industry as Creative Industries, namely advertising, architecture, arts and antique market, craft, design, designer fashion, film and video, interactive leisure software, music, performance arts, publishing, computer software and computer services, and television and radio. An update in the Creative Industries Mapping Document further defined Creative Industries as 'those activities which have their origin in individual creativity, skill and talent and which have the potential for wealth and job creation through the generation and exploitation of intellectual property' (Department for Culture, Media and Sport, 1998). This is a definition that focuses on economic efficiency and social benefit that is drawn from property rights and employment opportunities (Hong Kong Arts Development Council, 2000). Similar steps were taken in Singapore, Taiwan, Hong Kong and China.

The importance of the Creative Industries is growing among modern post-industrial knowledge-based economies. Apart from the high growth rate's contribution to economic wealth, the Creative Industries also foster cultural identity and social solidarity (HKADC, 2000) as well as acting as a catalyst for urban regeneration (Jones, Comfort, Eastwood & Hillier, 2004). Moreover, the Creative Industries are not subject to the traditional diminishing of marginal returns, and they do not consume large amounts of material resources as the cost of production. The adoption of the Creative Industries for future economic development can resolve problems of high energy consumption, pollution production and the low level of value added in manufacturing industries (Wu, 2006).

Chinese society is progressing towards modernization. The development of Creative Industries is an important step in the process of China's modernization and development. The primary reasons are:

1. The average annual income of a household in China has reached the level of USD 3,000. It signifies that people will demand more than daily necessities, want to have higher standards of entertainment and various kinds of cultural items to consume; and
2. Economic globalization means that China's industries are both directly and indirectly drawn into intense global competition.

In order to increase international competitiveness and maintain continuous growth, China should re-adjust the industrial structure and national policy in industrial development. Creative industries play a key role in enhancing China's core competitiveness and raising the added value of products and services. Moreover, it can also provide sustained growth without over-exploiting the natural resources and damaging the environment.

THE RAPID DEVELOPMENT OF ANIMATION EDUCATION IN CHINA

The vigorous support of the government in developing cultural and creative industries in China made many animation production companies move on from the role of original equipment manufacturer (usually known as OEM, indicating production work which is released under the name of the company commissioning the work). They instead turned part of their production scale into producing original titles. Moreover, many new animation production houses were formed to develop their own titles. Many government and private-funded animation bases also developed to provide support and facilities for animation productions (Wang, 2012). All these developments, as well as the support of government in expanding animation education, created a sudden boom in animation education development in China.

In 2000, there were only two education institutes providing animation education programs in China. By 2009, 1,279 higher education institutions were running 1,877 bachelor degree animation programs in China (Li, 2010). Owing to the different backgrounds and focuses of these institutes, they were providing animation education programs carrying distinct characteristics of their institutes. Institutes providing animation education courses mainly belonged to four types of background, namely arts, art and design, film and television communication, and computing. Most of them developed animation programs based on their existing resources. Some education institutes emphasized art and design aspects while some focused on the technology and computer skills (Ye, 2010). However, most of these programs saw discipline-specific and practical skills as the most important element of training in their curriculum.

ISSUES IDENTIFIED IN ANIMATION EDUCATION

In recent years, the rapid development in the animation education sector showed success in the sense of producing thousands of graduates. However, many studies indicated that the rapid increase in the number of animation education programs created many problems (Pan, 2012; Wang, 2007). The major concerns were in the following three aspects.

Inadequate Number of Qualified Teachers

One of the key concerns was the inadequacy of the resources available to satisfy the rapid expansion of animation education programs. Owing to the historical development of the animation education system, there were only a few education institutes providing suitable teachers in China. The booming demand for animation educators over the last ten years or so created a major shortage of qualified animation teachers (Sun, 2005; Tao, 2010). Qualified teachers should process both academic understanding and practical knowhow in animation work, as well as have good teaching skills. However, the rushed development of new animation programs and the over-expansion of the student intake in the last few years created an insufficient supply of qualified teachers. Fresh bachelor graduates from animation disciplines might have been employed to take up teaching in the newly operating animation programs (Xu and Cui, 2010). Without support from other senior animation faculty, their ability to function as professional animation teachers was far from clear.

Curriculum Does Not Match Market Demand

Owing to the intensive support of the government in developing new animation programs, many higher education institutes implemented animation programs irrespective of whether they had corresponding resources or not (Sun, 2005). Moreover, there was no agreed standard in the animation curriculum, some of the program managers might not have adequate knowledge and experience in animation education, and the developed program structures and program objectives were not well defined (Wang, 2012). On the other hand, some of the higher institutes developed their own curricula based on existing resources instead of market demand.

Focus or Diversify in Discipline Knowledge

Some scholars indicated that some of the animation programs were not focused specifically enough on animation study (Xu & Cui, 2010). They claimed that these programs were too broad in their areas of study and produced graduates who could not take up animation work independently (Sun, 2005). This phenomenon might be due to the presence of inadequately qualified teachers in the programs, who substituted other related topics for animation subjects. On the other hand, the programs might purposely work in this way to allow a wider choice of job opportunities for their graduates.

KNOWLEDGE GAP IN ANIMATION EDUCATION RESEARCH

In 2013, a search with the keyword "動漫教育" (animation education) in the China Journal Full-text Database caused a list of 423 articles to pop up with 201 articles examined issues related to the development of animation education in China. Most of them pointed out the various problems related to the rapid expansion of animation education over the last ten years. They covered human resources issues as well as curriculum development issues in the system. Some of them briefly mentioned the importance of creative thinking and creativity in animation education. However no article covered the nurturing of creativity in animation education.

In the China Animation Yearbook 2010, a chapter "中国动画高等教育发展报告" (Report on animation education in higher education development in China) showed a recent survey on the "动画教育面臨辦學困難" (Problems encountered in animation education) (Li, 2010). It listed 12 categories of problems identified with scores. The 12 categories prioritized in the order of importance were:

1. Lack of high level professional faculty,
2. Teachers' lack of practical industry experience,
3. Recession in the animation industry,
4. Low salaries of the faculty which prohibited them from focusing on teaching,
5. Inadequate professional facilities,
6. The weakness of practical study in the curriculum,
7. Lack of communication with the industry,
8. Lack of teaching materials like textbooks and audio visual materials,
9. Inadequate support from the institute,
10. Substandard quality of the applicants,
11. The fact that the program and curriculum structure were not well planned, and
12. The inadequacy of teaching space.

Among the 12 issues identified, none of them addressed the issues of creativity nurturing.

However, the same report also showed another survey, on the "动画专业学生需要的能力得分排序" (Ranking of the ability required for the animation major students). It listed 13 categories of abilities, and the priority in importance were:

1. Creative thinking,
2. Professional animation knowledge,
3. Drawing skills,
4. Cultural understanding,
5. Aesthetic ability,

6. Computer skills,
7. Ability to acquire knowledge,
8. Teamwork ability,
9. Independent work ability,
10. Expression and communication ability,
11. Ability to adapt to the environment,
12. Organization management ability, and
13. Foreign language level.

It was a surprise that "creative thinking" was ranked at the top of the list and the average score was 11.73, which was far higher than the next item on the list, "professional animation knowledge", with a score of 9.96.

From the above survey, it is apparent that animation educators really saw the importance of the type of creative thinking that is a key part of creative ability in animation education. This result shows the need of the cultural and creative industries: a demand for creative talent. However, the survey did not show awareness of there being any issue in the nurturing of creative thinking or creativity in animation education. The author of the article pointed out explicitly that the higher education institutes consider creativity supremely important in animation education, but at a certain level, lots of higher education institutes were relatively single-focused and did not have a systematic approach in recognizing principles in animation education. In addition, the author also indicated that the issue was how to actualize the nurturing of "creative thinking" in the program, and the fact that the curriculum needs in-depth exploration (Li, 2010). This comment showed a major dilemma in animation education in the nurturing of creativity. Most animation educators considered creativity important, but no research has attempted to take a closer look at this issue. This exposes a knowledge gap in this particular area, and this what this study aims to bridge.

THE IMPORTANCE OF CREATIVITY IN EDUCATION

The British Government set up a taskforce in 1998 to review creative and cultural education. The result was a report, All Our Futures: Creativity, Culture and Education, that led to an intensive debate on creativity education. Sir Ken Robinson, Chairman of the taskforce, challenged the traditional education system, saying: "The educational reforms really needed now are actually being held back by the attitudes to education that many policymakers learned when they went to school – 20, 30, 40 years ago. Many seem to believe the way to the future is simply to do better than what we did in the past. The truth is we need to do something completely different for today's students" (NACCCE, 1999, pp. 2). Education institutes should not limit themselves to knowledge transfer, but must also promote flexibility and openness for our students. Fostering creativity can help our students meet the challenges of the future (Cropley, 2001).

The NACCCE report indicated that education worldwide faced unprecedented challenges of many types: economic, technological, social and personal. A rapid change of the nature of work from the traditional industrial model to one based on knowledge and information called for new type of workers. The expanding creative industries also have a demand for young people who can generate new ideas. The new pattern of work has a demand for young people to switch occupation and location much more than the past. The rapid development of technology has already dominated all areas of our lives. It offers tremendous opportunities for people to develop and transform, but on the other hand it might also be harmful in social, emotional and imaginative development. Both economic and technological changes are transforming the social landscape. These changes will affect the balance of relationships and responsibilities between generations. Traditional academic quality will not satisfy the demands of employers who emphasize powers of creativity, of communication, of empathy and adaptability. Creative abilities are needed to meet the challenges of economic, technological, social and personal development (1999). Nowadays, employers demand people who 'can adapt, see connections, innovate, communicate and work with others' (NACCCE, 1999, pp. 14). Moreover, fostering creativity does not conflict with traditional education. Some scholars see creativity seeming to supplement conventional intelligence in promoting school performance (Sternberg & Lubart, 1995).

CREATIVITY AND CREATIVITY RESEARCH

Research into the nature of creativity has only a short history and there is no commonly accepted approach to deal with this topic. Moreover, different approaches on the definitions, conceptualizations, domains and disciplines make the study more complex. Creativity is difficult to define. Czikszentmihalyi described it as follows: "Creativity was recognized to be a complex, contested concept that is poorly theorized, as acknowledged in many accounts" (1999, pp. 313). Guilford (1971) saw "creativity" as an ambiguous word. Child (1973) suggested that there does not exist a clear, unambiguous and widely accepted definition of creativity. Stein (1953) proposed that ''creation'' is an activity which brings some things which are ''new'' and which have ''utility''. Sternberg and Williams (1996) also had a similar thought, that creative ideas are both novel and valuable. Boden defined creativity is "the ability to come up with ideas that are new, surprising, and valuable" (Boden, 2004, p. 1). Most of the scholars agreed that creativity involves thinking that is aimed at producing ideas or products that are relatively novel and compelling. (Kaufman & Sternberg, 2006).

The National Advisory Committee on Creative and Cultural Education defined creativity as "imaginative activity fashioned so as to produce outcomes that are both original and of value" (NACCCE, 1999, pp. 30). The definition of creativity proposed by the NACCCE was based on the understanding of the nature of creative processes and this indicative definition aimed to serve for education purposes. Creative processes always involve:

1. Thinking or behaving imaginatively;
2. The imaginative activity be purposeful;
3. The activity must generate something original; and
4. The outcomes must be of value to the purpose (NACCCE, 1999).

CATEGORIES OF CREATIVE MAGNITUDE

Creativity can be realized as creative expression (a product) or creative experience. These creative outcomes are different in creative magnitudes as well as methods of measurement. Differentiating the categories of creative magnitude helps us to understand the various approaches of creativity research.

The categories of creative magnitude address the differences of internal and external frames of reference of creativity. They are commonly divided into four categories, namely:

1. Big-C,
2. Pro-c,
3. Small-c, and
4. Mini-c.

Big-C refers to the eminent creative expressions that are generally recognized. Works of great poetry or substantial invention are considered to be in this category. These are eminent levels of creative expression, like Emily Dickinson's Poetry, John Coltrane's Jazz or Sigmund Freud's Psychology. Pro-c refers to creative expressions that exist at a professional level but do not yet attain, or never attain, the eminent level. Small-c refers to the objective creativity of everyday life. These creative expressions and experiences can be accessible in our daily life like creating a novel recipe that produces something which is enjoyed by family and friends. Mini-c refers to the more subjective creativity of everyday life including personal, internal, or mental forms of creativity.

RESEARCHES ON STUDENTS' CONCEPTION OF CREATIVITY

Many scholars suggest that the current education systems and approaches cannot meet the demands of today's fast-changing economy and society. They feel that education institutes should put extra effort into promoting creativity education. However creativity is a relatively complex concept, and successfully fostering creativity in education demands a clear understanding of what creativity and creativity education are (Jackson et al, 2006).

A study was carried out by Margaret Edwards, Chris McGlodrick and Martin Oliver (2006) to explore teachers' perspectives on creativity at Liverpool John Moores University and University College London in the period of 2002 to 2004. The research used a semi-structured approach and featured interviews with 32 academic staff from a range of disciplines and experiences. The general conclusion was that creativity was not a well theorized concept. Though some participants could indicate their understanding of the complexity of the concept, most of the others had problems and confusion with it (2006).

Another study was conducted in the same period, in the above-mentioned universities, with the participation of 25 students from various disciplines. Students were invited to participate in interviews and asked to explain their thoughts about creativity and to describe creative people or things. The result of the study showed that students had difficulty in explaining what they thought creativity was, and they were incapable of providing a coherent and integrated explanation of the term creativity (Edwards, McGoldrick & Oliver, 2006).

These studies actually showed that teachers and students generally have a confused conception of what creativity actually is. Although some of the participants might show a better understanding of the nature of creativity, the overall indications were that participants could not give a clear and coherent explanation of what creativity was.

RESEARCH APPROACHES AND DESIGN

This research project aims to explore how teachers and students perceive the meaning of creativity in education. Qualitative research puts the emphasis on discovery and description. This study employed qualitative research approach to collect, analyze and interpret data and present descriptive research findings, which would enable the bridging of the knowledge gap in this particular area.

Eighteen teachers and 30 students from animation programs of 10 universities in Beijing were invited to participate in the study. In a qualitative research approach, the selection of participants in a study is based on purposive sampling (Merriam, 1998). Unlike the random sampling techniques that are based on statistical probability theory, purposive sampling aims at identifying and selecting ''information-rich cases'' for study (Bloomberg and Volpe, 2008). Owing to the limited resources, the research focused on the animation education in a single city – Beijing. Beijing is the capital of China and the most influential city in China. It was also the earliest area to develop policies related to the creative industries, and most of the well-established animation schools are located in Beijing. In 2011, there were about 98 higher education institutes in Beijing, of which 36 provided animation education programs. Owing to the study's focused on creativity education, all the institutes providing vocational training were excluded. Ten institutes providing bachelor degrees or above were included in the research sample.

Data was collected using face-to-face semi-structured interviews for teachers and the corresponding final year supervisees. Each institute identified two teachers and four final year students to participate in the interviews. Two pilot interviews were conducted prior to the actual interviews and the questionnaire was improved based on the deficiencies identified in the pilot interviews. The study successfully interviewed 18 teachers and 30 students. All the interviews were then transcribed into text and input to the qualitative data analysis software for further analysis.

Qualitative data analysis software MaxQDA was used for the coding process. Multi-pass coding was performed to identify and refine the key codes and the conceptual framework that derived from the literature. The coding process helped to identify and note the common patterns in the data. Several data summary tables were produced to address the research questions.

RESEARCH FINDINGS AND DISCUSSION

The questionnaires for both teachers and students included direct questions on the interviewees' interpretation of the objectives of the program and their understanding of the term creativity. Some more follow-up questions on this point have been asked in most of the interviews. However, very limited information related to creativity was collected from the interviewees, and most of them replied with very vague statements. Below is a summary of the findings in the study.

Purpose of Animation Education Provides Vocation Training

During the data collection period, only a few programs provided some set of documents about their programs or program curriculum. None of the programs published sufficient information in their program websites to gain an understanding of their program objectives. However, the lack of official information on the individual programs did not affect the results of the research. Since the teachers had direct daily contact with the students, their influence on the students was immediate and substantial. How they saw the program and how they interpreted the program objectives would directly shape the actual program learning outcomes. Related questions on the purpose of the program were asked and the following results were identified.

The findings in the research show that the majority of teachers suggested that the main purpose of animation education was to train professional animation talent with production knowhow and practical skills. It shows that the interpretations of teachers and students are quite consistent in their understanding of the program objectives. Although most of them did not explicitly mention the term vocational training, many teachers implied in some kind of expression that they wanted to ensure their graduates should be readily employable and the program should prepare students for immediate jobs. Although a few teachers also emphasized providing a holistic education for students, the overriding concern of all participants was to provide sufficient knowledge and skills for the students to fulfill job requirements.

The expectations of the participants in animation education can be explained by current government policy. Since the government evaluates program performance on the graduates' employment rate, it creates a direct influence on curriculum development which aims to produce graduates who can fit the immediate demands of industry. Since the animation industry is in a transformational stage, most studios are still producing primarily OEM (original equipment manufacturer) work and a relatively small amount of original work, so the industry maintains a heavy demand for skillful animation workers, in a purely practical sense. On the other hand, some newly transformed OBM (original brand manufacturer) companies still retain the original OEM concept in their production. The industry people do not understand the value of creativity in the development of their business, so most of the time they undermine the importance of recruiting creative talent. The overall perception of these companies is that they maintain a small number of creative people in the top position and employ a large number of skillful animation workers to tackle the labor-intensive animation productions.

Owing to the industry demand for skillful animators instead of creative talent, the teachers perceive that knowledge acquisition is more important than the nurturing of creativity for students. Since the industry only demands a limited number of individuals with creative talent in top positions, many teachers might not see the importance of fostering creativity in animation education.

The current demand from the industry is for skillful animation workers who can fulfill the skilled staff member requirements for production work. Animation companies only demand a limited number of people with creative talent in top positions. The higher education institutes that provide animation education have to meet the industry demand and focus on training skillful animation talent. Fostering animation education is not their immediate concern.

Confusion on the Conception of Creativity

First, no student provided any definition for the term creativity. A majority of students suggested that creativity was a term which applied to something which included an aspect which was novel or differed from standard items. A few students also emphasized the point that ideas could only be considered as creative if they were expressed and realized.

Second, most of the teachers mentioned one or more subjects in their programs for nurturing students' creativity. However no details were provided by the teachers as to exactly how these subjects help in developing students' creativity. Moreover, no common pattern could be identified from these suggested subjects.

Third, some teachers considered that the curriculum for teaching creativity should involve some kind of teaching methodologies embedded in daily classes instead of some specific subjects in creativity training.

Fourth, a majority of students mentioned some subjects that they considered helpful for nurturing creativity. These subjects include taught subjects such as visual aesthetics, character design, body language and scriptwriting. Almost half of the students suggested that scriptwriting was helpful in developing their creativity. Some of the students also mentioned that production subjects were most helpful in developing their creativity, because these subjects allowed them to create some sort of story and characters, and they could also deal with practical problems.

Fifth, in nurturing students' creativity, a criticism-free environment was very important for them to generate creative ideas. A majority of the teachers claimed that they would respect students' decisions and should not force students to adopt their suggestions in completing the students' projects. Moreover, other teachers had very diverse recommendations for nurturing students' creativity. Some teachers suggested students absorb more knowledge, and develop independent thinking ability, aesthetic appreciation, synthetic ability, self-confidence, association ability, analytical ability, and multidisciplinary knowledge. Some teachers also suggested that it was important to help students find what excited them, and what inspired and motivate them in exploration.

Sixth, students did not express any unified idea on the way teachers helped them develop their creativity, but mentioned various ways that teachers encouraged them to be more creative. These included motivating and encouraging them, providing them with a supportive environment that allowed exploration, widening the scope of students in both cognition and knowledge, requesting students to look at things from different perspectives, and inspiring students by using different teaching approaches.

The major findings in this study indicated that some teachers believed "novelty" and "appropriateness" were two important elements of creativity, but most others had confused conceptions of creativity. It is interesting that students had more comments on their understanding of creativity. Apart from novelty and appropriateness, students also suggested that creativity was something that was personal, involved freedom from the routine, could be incremental, and was common to all, etc.

As Chinese culture has traditionally always valued collectivism more than individualism, the nature of creativity that emphasizes questioning assumptions and risk-taking might not always be appreciated. The cultural influence might indirectly make schools and teachers undermine the importance of creativity in education and thus they never pay serious attention to the meaning of creativity. According to Niu's (2006) study, the first peer-reviewed journal article on creativity in China was published in 1983. Before 1995, there were fewer than ten publications in peer-reviewed journals on creativity and none of them were empirical. This might further indicate the low level of perceived importance of creativity in China, at least until recently.

Teachers and students lack of a clear understanding of creativity is due to the fact that the importance of creativity has been undermined in education, in work and in the society. However, it was also found that teachers seldom directly declared that creativity was not important. Most of them saw creativity as important in animation education, but they could not give details on what they had done to promote creativity in the course.

The Cultural Influences on the Conception of Creativity

In comparing Chinese teachers and students with their UK counterparts on the understanding of creativity, it was found that both groups share the same confusion and hesitation about their conception of creativity. The lack of understanding of the conception of creativity in higher education institutes is an important issue if we want to foster creativity in the higher education system.

Although there have been many common suggestions from both teachers and students, there were some important descriptions from participants in the UK studies that were not seen among Chinese counterparts, such as, "an imagination of expression'', ''risk'', and ''independence''. On the other hand, the frequently mentioned ''appropriateness'' or ''value'' that emerged in the Chinese context was not mentioned by the UK counterparts.

As Kim (2007) suggested, creativity is influenced by various factors. Apart from the personal domain, creativity is also affected by other social factors and by culture. Owing to the difference in the cultural context, creativity would be understood and valued differently (Paletz & Peng, 2008). Culture can be understood as a set of conventionalized learned routines, and serves to coordinate social behaviors of individuals within that society (Chiu & Hong, 2006). The different conception of creativity by Chinese participants might be explained in part by their cultural background.

Culture can be defined as "an historically transmitted pattern of meanings embodied in symbols, a system of inherited conceptions expressed in symbolic forms by means of which men communicate, perpetuate, and develop their knowledge about and attitudes toward life" (Geertz, 1973, pp. 89). Niu and Sternberg (2003) proposed that in cross-cultural studies, the Core Societal Values should be studied in order to understand the differences between social groups with respect to individualism and collectivism. Western societies that favor individualism tend to emphasize an independent perspective of the self. Behavior and value judgments mainly make reference to the individual's internal thoughts instead of his or her relationship with others or with society at large. As Asian cultures tend to value collectivism over individualism (Kim et al, 2011), people in Eastern societies incline to favor an interdependent perspective of the self and try to become part of various personal relationships. The societal values, as suggested by Niu and Sternberg (2003), shape individual creativity through:

1. Influencing the understanding of creativity, and
2. These values further influence individual creative performance.

Moreover, Confucianism dominates the development of Asian cultures (Chaves, 2002), and serves as the point of reference for the core values, traditions, ethical and moral foundations of Chinese society (Kim et al, 2007). A Confucianist world view tends to prioritize the acquisition of societally-ordained "correct" knowledge on education and limits initiatives in generating new ideas (Ho & Ho, 2008; Martinsons & Martinsons, 1996). This might explain why the Chinese teachers and students interviewed in this study tended to emphasize knowledge acquisition and appropriateness in their program objectives and conceptions of creativity.

The Government Policy in Preventing Foreign Cultural Influences

The fast economic development of China in recent decades has already turned it into a manufacturing giant. However, China still relies heavily on its low cost of labor to attract outsourced production from companies of developed countries. This foreign investment enterprise will move away from China if the favorable factor of low labor costs disappears, as is already happening. In order to maintain international competitiveness, China has explored some other opportunities to sustain its economic growth.

Creative Industries are fast growing industries that 'provide job opportunities, create wealth, produce consumer goods and services for local and overseas market, enable growth in overall consumption, promote social solidarity as well as export cultural influence' (HKADC, 2000). The key success of the creative industries is built upon the deployment of creativity through the exploitation of intellectual property (DCMS, 1998). Democratic societies do not consider creativity to be politically problematic, but this might not be necessarily be the case for authoritarian regimes (Wang, 2004). As suggested by Keane, "in a political system that views culture as a public resource, the momentum of market forces raises issues of 'national cultural security'" (Keane, 2013, pp. 3).

Developing cultural and creative industries can help China to maintain competitiveness in the economic development. After the reform and opening of the economy of China in the 1990s, many state-owned media and cultural institutions have been privatized. In order to protect these newly developed cultural industries from foreign competition, instead of using some dogmatic interventions, the national government promoted cultural industries with creativity and innovation as a national strategy to avoid the influx of foreign cultural products and influences (Li, 2011). Some scholars suggested that adopting the term creative industries for the cultural industries in China was done mainly "to evade the dogmatic structures of the state cultural policy" (Keane, 2013, pp. 3).

As creativity is always associated with ''non-conformists'', ''eccentricity'', and ''abnormal'' situations (Handy, 2004; Guastello, 2009), it is often associated with deviation from a given national culture, and thus counter to the policy of many authoritarian regimes in emphasizing conformity. China, as a socialist society which emphasizes collective development and social stability, thus possesses some innate internal barriers to the development of a creative culture. Above all, the Chinese government still exercises a tight control and censorship on content shown in television broadcasts. This control policy also illustrates the intention of the government's conditional support to the development of creativity in the cultural and creative industries. It is fair to argue that the national government's interest in the creative industries is not aimed at producing more creative products but maintain a non-dogmatic approach to prevent the influx of foreign cultural influences.

The Demand for Creativity and Innovation of the Creative Communities

Owing to rapid changes in society, labor intensive jobs are being lost to lower-cost regions, and the developed economies facing the challenge of achieving economic and social stability must consider the importance of knowledge and innovation in their national strategies. In order to achieve this goal, governments should facilitate learning and also teach people how to learn (Hodgson, 2000). Nurturing a learning society with flexibility and creativity can help promote sustainable economic growth and social stability. China, as one of the biggest manufacturing countries, is also facing the same challenge. The rise of labour costs is driving many manufacturing jobs to regions like Vietnam, Thailand and the African countries. The urgency of change to a more innovative and creative economy is high in China.

Most of the developed countries have gone through a transformation from being industrial economies to being knowledge economies. Success in a knowledge economy is built upon human capital. As Mahbubani (2008) suggests, Asian minds have been underused for centuries and they are now exploding with creativity. He also considers that China, even though it is not a democratic society, will play a leading role in this transformation.

Keane (2013) quoted the argument of Collins (1998) in expressing the view that creativity has been demonstrated throughout Chinese history in the form of great inventiveness, scientific innovation and aesthetic achievement, and was not simply imported as the ''West pre-packaged'' creative industries in 2004. Although traditional Confucian culture, which emphasizes harmony and reciprocity, might create certain potential conflicts in the development of creativity in the way that it developed in Western cultures, with the inevitable growth of globalization and modernization, the demand for creativity and innovation provides the opportunity for Chinese society to develop a special type of creativity which would be a fusion of Confucianism, socialism and capitalism.

As Keane (2007, 2013) suggested, China has gone through six stages of development in its creative economy:

1. Standardized production,
2. Imitation,
3. Collaboration,

4. Trade,
5. Clusters, and
6. Creative communities.

The six stages illustrate how China is moving from being a simple material culture to being an intangible, complex and fragmented digital economy. Keane also pointed out that the last stage of creative communities, which facilitate creation and recreation, is considered as the most revolutionary and potentially innovative stage in the timeline. The dialogue with both teachers and students in the interviews largely indicated their demands and expectations for creativity in education, thought they did not have a clear conception of the term or well-defined ways to actualize it in the education process. The present state of creative communities in China signals a demand for, and appreciation of creativity in societal development, irrespective of the control of the state. This bottom-up approach of seeing the needs in the development of creativity can facilitate the growth of the creative culture and promote a better conception of creativity among people. As Keane says, the Chinese government should "focus… on stimulating creative education rather than ensuring cultural maintenance" (Keane, 2013, pp. 47).

CONCLUSION

As many scholars have said, if teachers do not have a clear conception of what creativity is, it is very difficult for them to foster creativity in their teaching. A study was carried out by Margaret Edwards, Chris McGlodrick and Martin Oliver in exploring teachers' perspectives on creativity at Liverpool John Moores University and University College London in 2002. The general conclusions were that creativity was not a well-theorized concept, and although some participants could indicate their understanding of the complexity of this concept, most of the others had problems and confusion in articulating the concept. A similar situation happened in this study. Many students also found it difficult to articulate their understanding of creativity. It has been identified in the findings that most of the participants did not have a clear conception of what creativity was. Although some of them related novelty to creativity, and a few of them also talked about appropriateness as one of the factors, the overall conception of what creativity actually is, was very confused. There are multiple reasons that contribute to this situation.

As Chinese culture always values collectivism more than individualism, the way creativity emphasizes the questioning of assumptions and the taking of risks might not always be appreciated. The cultural influence might also affect schools in undermining the importance of creativity in education. From previous findings, It have been determined that neither teachers nor students could offer a clear articulation of their conception of creativity. This might largely be due to the fact that they never pay serious attention to the meaning of creativity. According to Niu's (2006) study, the first peer-reviewed journal article on creativity in China was published in 1983. Before 1995, there were fewer than ten publications in peer-reviewed journals on creativity and none of them were empirical. This might further indicate the low level of perceived importance of creativity in China, at least until recently.

Teachers and students do not have a clear understanding of what creativity is because the importance of creativity has been undermined in education, in work and in society. However, the study also found that teachers seldom directly declare creativity to be unimportant, and most of them see creativity as being important in animation education. The expectations about creativity in education from both teachers and students come mainly from the demands of societal development.

The fast development of the cultural and creative industry sector in recent years shows that China has set a goal to promote creativity. However, it is impossible to foster creativity in education if the teachers do not have a clear conception of what creativity is. It is apparent that educators see the value of creativity but they might still have a confused conception of what creativity is. In order to develop a better understanding of creativity, creativity researchers should work closely with educators in finding an appropriate meaning of creativity and build a better understanding on the nature of creativity in a Chinese context.

FUTURE RESEARCH DIRECTIONS

Although the focus of this research is on the level of higher education, the study of the nurturing of students' creativity at various levels, including primary and secondary education, also deserves immediate attention. The establishment of the National Advisory Committee on Creative and Cultural Education and the release of the NACCCE report focus mainly on this concern. In order to address the issue of nurturing students' creativity in education, both top-down and bottom-up approaches should be considered. Joubert suggested "a bottom-up approach can ensure a feeling of ownership toward the initiative" (2001, pp. 32), so the understanding of teachers' conception of creativity and the way they promote teaching for creativity shall be an important direction in the future study.

REFERENCES

Beghetto, R. A. (2007). Does creativity have a place in classroom discussions? Prospective teachers' response preferences. *Thinking Skills and Creativity*, *2*(1), 1–9. doi:10.1016/j.tsc.2006.09.002

Bloomberg, L. D., & Volpe, M. (2012). *Completing your qualitative dissertation: A road map from beginning to end*. Sage Publications.

BoF. L.JingY. Y.LiuW. (薄芙丽, 景怀宇, & 刘玮) (2009). 探索实践教学新路 优化人才培养规格——关于中国动画教育改革的设想. 装饰. 6, 042.

Boden, M. A. (2004). *The creative mind: Myths and mechanisms*. Psychology Press.

Child, I. L. (1972). Esthetics. *Annual Review of Psychology*, *23*(1), 669–694. doi:10.1146/annurev.ps.23.020172.003321

Chiu, C., & Hong, Y. (2006). *Social psychology of culture*. New York: Psychology Press.

Collin, R. (1998). *The Sociology of Philosophies: a Global Theory of Intellectual Change*. Harvard University Press.

Cropley, A. J. (1992). *More ways than one: Fostering creativity*. Ablex Publishing.

Cropley, A. J. (2001). *Creativity in education & learning: A guide for teachers and educators*. Psychology Press.

CsikszentInihalyi, M. (1999). Implications of a systems perspective for the study of creativity. In R. J. Sternberg (Ed.), Handbook of creativity (pp. 313-335). New York: Cambridge University Press.

Dawson, V. I. (1997). In search of the wild bohemian: Challenges in the identification of the creatively gifted. *Roeper Review*, *19*(3), 148–152. doi:10.1080/02783199709553811

Department for Culture. Media and Sport. (1998). Creative industries mapping document. Author.

Edwards, M., McGlodrick, C., & Oliver, M. (2006). Creativity and curricula in higher education: Academics' perspective. In N. Jackson, M. Oliver, M. Shaw, & J. Wisdom (Eds.), Developing creativity in higher education: An imaginative curriculum. Routledge.

Geertz, C. (1973). *The Interpretation of Cultures: Selected Essays*. Basic Books.

Guastello, S. J. (2008). Creativity and personality. In T. Richards, M. A. Runco, & S. Moger (Eds.), *The Routledge companion to creativity* (pp. 267–278). Routledge. doi:10.4324/9780203888841.ch22

Guilford, J. P. (1970). Creativity: Retrospect and prospect. *The Journal of Creative Behavior, 5*, 77–87. doi:10.1002/j.2162-6057.1971.tb00877.x

Handy, C. (2004). *The new alchemists: How visionary people make something out of nothing*. London: Hutchison.

Ho, D., & Ho, R. (2008). Knowledge is a dangerous thing: Authority relations, ideological conservatism, and creativity in Confucian heritage cultures. *Journal for the Theory of Social Behaviour, 38*(1), 67–86. doi:10.1111/j.1468-5914.2008.00357.x

Hodgson, G. (2001). Socio-economic consequences of the advance of complexity and knowledge. *Voprosy Economiki, 8*.

Hong Kong Arts Development Council. (2000). Introduction to Creative Industries: The Case of United Kingdom and Implementation Strategies in Hong Kong. Hong Kong: Author.

Jackson, N., Oliver, M., Shaw, M., & Wisdom, J. (Eds.). (2006). *Developing creativity in higher education: An imaginative curriculum*. Routledge.

Jones, P., Comfort, D., Eastwood, I., & Hillier, D. (2004). Creative industries: Economic contributions, management challenges and support initiatives. *Management Research News, 27*(11/12), 134–145. doi:10.1108/01409170410784644

Jubert, M. M. (2001). The art of creative teaching: NACCCE and beyond. In A. Craft, B. Jeffrey, & M. Leibling (Eds.), *Creativity in education* (pp. 17–34). London, UK: Continuum.

Kaufman, J. C., & Sternberg, R. J. (Eds.). (2006). *The international handbook of creativity*. New York: Cambridge University Press. doi:10.1017/CBO9780511818240

Keane, M. (2013). *Creative industries in China: Art, design and media*. John Wiley & Sons.

Kim, K. H. (2007). Exploring the interactions between Asian culture (Confucianism) and creativity. *The Journal of Creative Behavior, 41*(1), 28–54. doi:10.1002/j.2162-6057.2007.tb01280.x

Kim, K. H., Lee, H. E., Chae, K. B., Anderson, L., & Laurence, C. (2011). Creativity and Confucianism among American and Korean educators. *Creativity Research Journal, 23*(4), 357–371. doi:10.1080/10400419.2011.621853

Li, H. (李海玲) (2010). Report on the Higher Animation Education in China (中国动画高等教育发展报告). In China Animation Yearbook 2010 (pp. 469-479). Communication University of China Press.

Mahbubani, K. (2008). *The new Asian hemisphere: the irresistible shift of global power to the East*. PublicAffairs.

Martinsons, M. G., & Martinsons, A. B. (1996). Conquering cultural constraints to cultivate Chinese management creativity and innovation. *Journal of Management Development*, *15*(9), 18–35. doi:10.1108/02621719610146239

Merriam, S. B. (1998). *Qualitative research and case study applications in education. Revised and expanded from "Case study research in education"*. Jossey-Bass Publishers.

National Advisory Committee on Creative and Cultural Education. (1999). *All Our Futures: Creativity*. Culture and Education.

Niu, W. (2006). Development of creativity research in Chinese societies: A comparison of Mainland China, Taiwan, Hong Kong, and Singapore. In J. C. Kaufman & R. J. Sternberg (Eds.), The international handbook of creativity. Cambridge University Press.

Niu, W., & Sternberg, R. J. (2003). Societal and school influences on student creativity: The case of China. *Psychology in the Schools*, *40*(1), 103–114. doi:10.1002/pits.10072

Paletz, S. B. F., & Peng, K. (2008). Implicit theories of creativity across cultures: Novelty and appropriateness in two product domains. *Journal of Cross-Cultural Psychology*, *39*(3), 286–303. doi:10.1177/0022022108315112

Pan, J. (2012). *Report on the China Animation and Comic Education (中国动漫教育发展报告) 2010-2011*. China Social Sciences Press.

Scott, C. L. (1999). Teachers' biases toward creative children. *Creativity Research Journal*, *12*(4), 321–337. doi:10.1207/s15326934crj1204_10

Stein, M. I. (1953). Creativity and culture. *The Journal of Psychology*, *36*(2), 311–322. doi:10.1080/00223980.1953.9712897

Sternberg, R. J., & Lubart, T. I. (1995). *Defying the crowd: Cultivating creativity in a culture of conformity*. Free Press.

Sternberg, R. J., & Williams, W. M. (1996). *How to Develop Student Creativity*. ASCD.

Sun, L. (孙立军) (Ed.) (2005). The Situation of Local Animation Education (国内动画教育状况). China Animation Industry Annual Report 2005 (pp. 112-137). China: Ocean Press.

Tao, L. (陶琳娜) (2010). The Requirements of Teachers in Film and Television Education (影视动画教育对教师的要求). *Master*. 163.

Wang, J. (2004). The global reach of a new discourse: How fast can 'Creative Industries' travel? *International Journal of Cultural Studies*, *7*(1), 9–19. doi:10.1177/1367877904040601

Wang, K. (王可) (2012). Thinking of a few questions in the Higher Animation Education (关于中国高校动漫教育若干问题的思考). *Art in China*, (2), 115-116.

Wang, Z. (王正友) (2007). An In-depth Study of the Current Situation of China Animation Education (中国动画教育现状探析). *Education and Vocation*, (9), 103-104.

Westby, E. L., & Dawson, V. L. (1995). Creativity: Asset or burden in the classroom? *Creativity Research Journal*, *8*(1), 1–10. doi:10.1207/s15326934crj0801_1

Wu, Q. (2006). Creative industries and innovation in China. *International Journal of Cultural Studies*, *9*(3), 263–266. doi:10.1177/1367877906066872

Xiao, Y. (2011). An Analysis of the Current Situation in China Animation Education. *Journal of Beijing Union University*, *9*(3), 71–77.

Xu, C., & Cui, B. S. (2010). *A Discourse on Current Situation and Strategy of University Animation Education*. Heilongjian Science and Technology Information.

Ye, J. (2010). The Current Situation in China Animation Education, an Analysis of the Problems and Development Strategies. *Journal of Hubei University of Economics*, *7*(8), 174–175.

KEY TERMS AND DEFINITIONS

Animation Education: Animation education in China includes degree-awarding programs and non-degree-awarding vocational training programs. Animation education discussed in this chapter mainly relates to degree-awarding programs.

Collectivism: This is an ideology that emphasizes collective benefit over individual benefit when the two come into conflict.

Confucianism: A school of thought traced back to the Chinese philosopher Confucius. It is a way of thinking that constitutes the mainstream philosophy of the Chinese culture.

Creative Industries: The concept of "the Creative Industries" in the modern sense can be traced back to the establishment of the Creative Industries Task Force by the UK Government's Department of Culture, Media and Sports (DCMS) in June 1997. Creative industries aim to promote economic efficiency and social benefits drawn from intellectual property rights and employment opportunities.

Creativity Education: The early elitist view believes that creativity is a gift and can only be found in a few individuals. Most educational researchers and theorists participating in promoting creativity reject that elitist view. They believe everybody has a certain potential to be creative, and it is the responsibility of the school to foster the fullest development of this potential.

Creativity: Creativity is a complex concept and there is no commonly accepted definition in academic circles. But most of scholars agree that creativity involves thinking that is aimed at producing ideas or products that are both ''novel'' and ''appropriate'' to a certain purpose.

Individualism: This is an ideology that emphasizes individual rights and freedom.

Chapter 5
Chinese Students' Experiences in American MFA Programs in Digital Arts:
How to be Prepared and What to Expect

Edmond Salsali
Georgian Court University, USA

Rebecca Ruige Xu
Syracuse University, USA

ABSTRACT

In recent years, the number of Chinese students seeking advanced education in the United States has increased considerably. This paper attempts to compare Chinese students' expectations of the MFA program in digital arts in the U.S. to the actual contents offered by the graduate schools. It addresses the unique challenges they face when pursuing MFA degrees and discuss how they could overcome those challenges to successfully achieve their educational goals, as well as how the MFA programs can assist them in this endeavor. It explains that digital arts curriculums in the US are usually meant to help students achieve an individual approach to the field, while

DOI: 10.4018/978-1-5225-0911-0.ch005

developing their aesthetic and artistic sensibilities through conceptualization. But upon entering graduate schools, Chinese students usually lack proper knowledge of the theory, history, and philosophy of art, which are employed extensively in these programs to engage students in a multidisciplinary practice that emphasizes as much on the conception of the final piece as it does on its technical prowess.

INTRODUCTION

In recent years, the number of Chinese students pursuing MFA degrees in the United States has increased considerably. Since the goal of this book is to explore design education and promote creativity and innovation in Chinese education, the authors find it appropriate to focus their attention primarily on the creative subjects of art and specifically digital arts at the graduate level. Similarly to most other majors, Digital arts programs in the US have also witnessed a significant increase in the number of Chinese students. As an example, for the 2014-15 academic year, 80% of applicants to the Digital Arts graduate programs at Syracuse University were from China. Therefore, it becomes necessary to offer a glimpse of insight on how these MFA programs function.

This paper is an attempt to compare Chinese students' expectations of such American programs to the actual contents offered by the graduate schools. It wishes to address the unique challenges these students have to face and discuss how they could overcome those hurdles to successfully achieve their educational goals. Based on their many years of teaching digital arts and multimedia in American academic institutions, the authors will discuss the different ways in which the MFA programs can assist them in this endeavor, and how Chinese institutions can prepare them more appropriately for these challenges from an early stage, prior to moving to the US. The final goal is to assist these students in conducting more successful studies in their graduate programs.

BACKGROUND

Artists began experimenting with computers to create artworks since the 1950s. The first computer art show entitled Computer Generated Pictures was held at the Howard Wise Gallery in New York in 1965. This exhibition was strongly influenced by contemporary art movements and popular culture. Moreover, since many forms of digital arts heavily rely on interactivity and viewers' participation, art historians have often related their roots to the Dada and Fluxus art movements, in which audience participation was already important. As Christiane Paul (2008) mentions: "digital art did not develop in an art-historical vacuum, but has strong connections to previous art movements, among them Dada, Fluxus, and conceptual art. The importance of these movements resides in their emphasis on formal instructions and in their focus on concept, event, and audience participation, as opposed to unified material objects" (p. 11).

It wasn't until the arrival of personal computers that a significant rise in the number of different genres of computer related artworks became noticeable. Since then, curators have referred to these forms of arts as "computer arts," "new media arts," or "digital arts." They typically include computer generated imagery, computer animation, physical computing, creative computer programming, digital music, sonic art, visual effects, gaming, virtual reality, and other interactive art forms. With constant expansion of digital technologies, new art genres continue to emerge each day, including robotic art, bio art, intelligent art, etc.

Similarly to their traditional counterparts, all these digital art practices expand beyond simple adoption of new tools, to address expressive or conceptual creative objectives, and respond to the artistic queries raised by individuals. Therefore, the artists' responsibilities are not limited to learn the practical aspects of using the tools, which could be a tedious task on its own, but they also need to develop a deep understanding of the underlying conceptual art theories and concepts.

MFA Education in Digital Arts in the United States

In the U.S., digital arts curriculums are designed to foster experimentation and realization of new art forms. They encourage unique solutions and emphasize on individuality and self-expression. In contrast, Chinese art education stresses skill building and technical expertise, and focuses on hands-on practice. In China, most students expect to follow their professors' research pathways and are not accustomed with setting their own research criteria and conducting their own investigations. They often go through extensive art foundation trainings and exhibit technical proficiencies in various fields. But since they don't receive sufficient theoretical groundings, their key challenges in the US become conceptual development of their own artworks as well as intellectual analyses of various art forms. Their lack of knowledge about modern art history and digital art theories prevents them from achieving their full artistic and developmental potentials, and when they first begin their graduate studies in the US, they often experience a degree of confusion and misconception about how to be individual-minded, assertive, and analytical (Zang & Sternberg, 2012).

The term "digital arts" refers to various art forms that make extensive use of digital technologies and utilize them as the main ingredient of the creative process. These practices are nowadays rather established and feature distinguishable characteristics, including the use of computer hardware and software, interdisciplinarity, on-screen viewing, and real-time interactivity. Digital arts trainings in American graduate schools bring arts and sciences together and require the artist to be knowledgeable in both areas. They are usually meant to help students achieve an individual approach to the field, while developing their aesthetic and artistic sensibilities through conceptualization. They critically engage students to analyze their own work as well as others, and theoretically interpret the relationship between those works and their underlying technologies. Theory and history of art are employed extensively in conjunction to advanced sciences and technologies, in order to engage students in a multidisciplinary practice, which emphasizes as much on the conception of the final piece as it does on its technical prowess.

A Master of Fine Arts degree (MFA) is recognized by many institutions as a terminal degree for practitioners in visual and performing arts, which include Fine Arts, Design, Dance, Theatre, Film/Video, Digital Arts, and Creative Writing. In these disciplines, an MFA degree is required as a minimum qualification to apply for a full-time teaching position at the university and college levels.

As stated in the College Art Association's (2015) MFA standards in visual arts and design, "the MFA degree demands the highest level of professional competency in the visual arts and contemporary practices. To earn the MFA, a practicing artist must exhibit the highest level of accomplishment through the generation of a body of work. The work needs to demonstrate the ability to conceptualize and communicate effectively by employing visual language to interpret ideas. In addition, the MFA recipient must give evidence of applying critical skills that pertain to meaning and content, ultimately encouraging a comprehensive examination and critique of the function and role of art from a variety of views and contexts." In other words, students' artworks should show skillful practical executions and their dexterity to use various tools of the trend. But most importantly, they need to demonstrate their ability to conceptualize and communicate creative messages effectively. They should employ visual semantics and morphologies successfully in order to convey their artistic objectives to the viewers.

The MFA degrees in digital arts are artistic research and development programs that situate themselves in the framework of digital technology. These programs are dedicated to a wide range of practices in art media where the computer plays a central role in the executions. The media arts created by computers may comprise interactive installations, electronic objects and interfaces, moving images, 3D modeling and animation, tactical media, rapid prototyping, internet art, and panoply of other continuously emerging media procedures.

As mentioned earlier, these MFA curriculums are designed to help students discover an individual aesthetic and artistic vision. They provide a good balance between art history, critic and theory of art, and studio practices. Advanced technologies are explored as creative tools and agents, which can lead to the production of new meanings, philosophies, and artistic inventions, all of which are central to the contemporary art practices. Students are expected to develop a strong record of professional practice in a variety of contexts including exhibitions, screenings, public interventions, installations, performances, workshops, and gaming events, among other forms of public engagements.

From the above explanations, it becomes clear that digital arts are technology-related activities that are diverse and multidisciplinary, with outcomes that can be associated to a multitude of creative fields. Therefore, MFA programs in digital arts accept students with various educational backgrounds ranging from graphic design, multimedia, animation, video, and gaming, to more traditional visual arts areas such as painting, sculpture, and film. Individuals with prior trainings in many other arenas such as mathematics, physics, computer science, and architecture can also apply, as long as they provide appropriate artistic portfolios.

Backgrounds and Admissions of Chinese MFA Students

The Chinese government has already started to take constructive measures to pedagogically assist the growing number of Chinese students in American institutions. According to Jiang Xueqin (2011), deputy principal of Peking University High School, "with Beijing's official backing, and facing increasing demand from parents, public schools in China's largest cities are heading toward a two-track system—the traditional stream that prepares students for the national college-entrance examination (the gaokao), and a new one that prepares them for the SAT Reasoning Test and the Test of English as a Foreign Language, or Toefl. Some of China's very best public schools, where the smartest and wealthiest students congregate, will one day become international in focus." It would be advisable that undergraduate degrees in Chinese universities follow the same trend and orient their efforts towards preparing students who plan to pursue their graduate studies in the US in advance. The preparation should be twofold and address the two most prevalent needs that are currently noticeable among these students, overtaking language barriers and inspiring self-expression.

One of the biggest hurdles facing Chinese students upon their arrival to the United States, despite obtaining satisfactory Toefl scores, is their lack of proficiency in communicating in English and comprehending its various linguistic nuances. As an evident prerequisite to an effective American education, this encumbrance should be dealt with early on, while they are still in the mainland. Students who land in a program with college-level English and understand its different morphological codes, can capitalize on their efforts and focus mainly on the subject matter of their studies, rather than spending a considerable amount of time to learn the semantics of the language. "The main problem is not just to talk like a native speaker but to think like one" (Xueqin, 2011). Chinese institutions should acknowledge the epistemological needs for students who plan to study in the US at an early stage, in order to be able to prepare them suitably. The intellectual preparation for those students should be more closely related to their American counterparts, so that when they get to the US, they can more easily acclimate to the cultural transition.

In an article for the South China Morning Post, Mandy Zuo (2015) mentions that about 8000 Chinese students were expelled from American schools in 2014. She explains that the main reason for this failure has been their under preparedness at the time of enrollment, for which, she blames the poor Chinese academic discipline, which results in difficulty in adapting to Western schooling.

The second challenge that the majority of Chinese students encounter throughout their studies in the US is a certain degree of deficiency in self-expression. This problem is an important focus of this paper, since digital arts, like other art forms, heavily rely on individuality and personal style.

Along with a satisfactory Toefle score, digital arts applicants at a US university are required to submit a personal statement describing their main academic interests, and their reasons for choosing their degree of choice at the elected university. They also have to indicate their projected outcome from their graduate studies. A good statement should provide a clear understanding of the applicant's future goals and help faculty determine if he/she fits the program suitably.

In addition to this statement, the most important admission requirement is a sample portfolio, which is judiciously examined by faculty members during the admission process. MFA programs usually prioritize applicants whose portfolios illustrate a commitment to digital arts and exhibit evidence of creativity, individuality, and personal methodology. The technical qualities of the artworks are reviewed but not overwhelmingly emphasized. In general, Chinese students' portfolios tend to stress their technical achievements rather than their conceptual approach. As a result, many portfolios appear to be lacking in adequate originality and personality. We recommend that students envisioning to apply, build their portfolios accordingly and prepare artworks that showcase innovative ideas and distinctive characters, while limiting the number of pieces that only illustrate the applicants' technical abilities.

Academic Transition to Graduate Schools

Digital arts curriculums are designed to foster experimentation and realization of new art forms and help students develop their own and unique aesthetic visions.

These curriculums usually follow specific developmental paths. They start by guiding students through several processes of exploration. Consecutively, they provide grounds for experimentation, and finally they encourage students to define their own creative demeanor.

Graduate programs in computer arts in most universities are typically 3-year 60-credit MFA majors. During the first year, whilst struggling to get accustomed with a completely new culture, Chinese students have to furthermore confront the academic rigors of the U.S. graduate school for the first time. Linguistic deficiencies and lack of awareness of how the U.S. academic system functions may be among the initial hurdles to overcome. Students generally take a few skill-based and a number of theoretical courses that expose them to novel artistic practices and original creative thinking. While Chinese students generally display relatively strong technical skills, they often need to pay specific attention to their theoretical courses and conduct as much external research in related areas as possible.

The second year will be a continuation of creative explorations and experimentation within digital media and advanced technologies. The work load isn't necessarily more than what students are used to from their prior studies in China, but rather different. For example, students in the United States may take fewer courses but delve much deeper into the subject matter of each course than they might in China.

The third year is devoted to the final thesis, which requires a clear description, a thorough investigation, and a final realization. Through a series of studio and seminar courses, students are encouraged to consider the relationships between intent and content in the context of their work, and analyze their methodologies in writing. At this level, MFA candidates should demonstrate concrete capabilities in implementing digital technologies in their artworks. They should also display coherent and tangible knowledge of contemporary art theories, along with the ability to use this knowledge creatively in their own work. Expressivity and conceptuality take center stage and undeniably become imperative elements of students' exertions, and technicality becomes ancillary.

According to Susan Cain (2012), Chinese people are relatively introverted. They consider self-expression as a form of selfishness and avoid voicing themselves in public. They find assertiveness to be the result of aggressive and egocentric behaviors.

On the contrary, Western culture encourages independence and singular mindsets. It considers self-confidence and divergence of thoughts as a precondition to creativity and innovation. It strives for originality, not duplication. Bringing forth one's own point of view and adding one's personal opinion to the subject is considered more important than following someone else's ideas. Such an ideology doesn't negate the consideration of others' thoughts, but proposes to study and analyze a fusion of diverse ideas to eventually craft one's own perceptions and unique notions, in what Gaston Bachelard (1942) describes as a "humble and conscious narcissism devoid of pride" (p. 36). Therefore, Western mentality considers a digital artist's capacity of self-awareness as a virtue and a precondition to imagination and inventiveness.

Digital art education in China stresses on skill building and technical ability and arts curricula mainly focus on rote methods. When they first come to the United States, it usually takes some time for Chinese students to familiarize themselves with the American concept-oriented art education. Therefore, the earlier they become aware of the important theoretical areas they need to focus on, the faster they can efficiently integrate themselves in this educational system. This will ultimately lead them to produce works that meet their MFA criteria, while concurrently express their personal artistic objectives. To that effect, it is imperative to acknowledge the differences between American and Chinese ways of conceptualization. As Thomas Misco (2015) explains: "generally speaking, students from China subscribe to metaphysical defaults that are holistic, dialectical, collective, averse to confrontation and controversy, contextual, relational, harmony-oriented, and focused more on the social world. Students from the US generally demonstrate more attitudes of determinism, focus on the salience of individual things, individual human agency, individual mindedness, analysis, debating, feeling in control, and causal attribution" (p. 38). This analysis indicates that Chinese ideology draws extensively from sensitivity and feelings and could be endorsed under a mystical and idealistic Kantian perspective, while Western philosophy, which generally speaking relies mainly on analytical logic, could be categorized as a typically rational Hegelian mindset. Portraying the distinctions between oriental and occidental mentalities, Christine Buci-Glucksman (2001) defines the Asian outlook as a culture of ephemerality, impermanence, fluidity, and lightness. She goes on to clarify that these values, which for long used to have a negative connotation in the West, have eventually found their rightful position among occidental philosophies (p. 16).

Keeping an open mind towards western contemporary culture and philosophy and drawing inspiration from them, Chinese students in digital arts could utilize their strong sense-oriented ethos to create their own distinctive and original artistic styles. To achieve this goal, they will need to attain a thorough interpretation of both eastern and western philosophies and histories of art. At the same time, they should remember that relying exclusively on traditions will not lead to innovative ideas and creativity. If they wish to have an eye on the traditional elements of the Chinese culture, they should also modernize and use them in the context of the current art theories and practices.

Chinese Students' Expectations vs. the Actual Contents of American MFA Curriculums

Upon their entrance to graduate programs in the United States, Chinese students have specific anticipations, which often differ from what the curriculums might offer. The following list is a general comparison between some of those expectations and the actual contents of graduate programs:

1. **Professors' Supervision of Students:** In China, students anticipate versus their professors to lead their projects continuously and guide them closely through every step of their studies. In return, they follow their mentors' trails without reservation. Most graduate students in China are expected to pursue the professors' research pathways unconditionally. In the US, the affiliation between students and teachers, especially at a graduate level, is of a different nature. Students are expected to conduct their researches and execute their projects on their own. The professors are there only to observe, bring constructive suggestions, and advise. The final decisions and realizations are up to the students. The requirement of the majority of the programs is for students to conduct their own investigations on multiple subject matters of their choosing to eventually discover and determine a specific topic for their final thesis. Different professors oversee each of the steps of this process, and the concluding thesis is conducted under the supervision and guidance of the student's advisor. The aim of this procedure is to inspire students in the development of distinctive visions as artists and prepare them in setting their own standards not only throughout their academic studies but for the rest of their professional careers.
    a. Another benefit of this method is to prepare students to adapt more easily with the ever evolving field of new media. Since the required technology is sophisticated and changes continuously, students in this field need to have the ability to find the answer to their technical questions on the go and on their own. To realize their projects, students will often find themselves confronted to unpredicted situations necessitating new and innovative technics and creative approaches. With a clear understanding of how to conduct thorough research, students will be able to find the answers to their questions in a timely manner and effectually. It is important to note that Chinese students consistently demonstrate an excellent ability to adapt to new and rapidly shifting technological requisites of their fields. Fast-pasted adjustment to new situations can be perceived as a characteristic of the contemporary Chinese culture as a whole. As

Pierre Picquart (2011) explains: "it's all of China and its entire society that change on a monthly basis. A phenomenon that is not imaginable in occidental countries. China advances with an astonishing speed in every domain" (p. 151).

2. **Course Content and Purpose**: During their course of studies, one of Chinese students' main expectations is to improve their technical skills. But in the US, students are expected to have already attained a good level of practical ability by the end of their undergraduate education, and although graduate program curriculums provide ample hands-on practices, their primarily focus is students' conceptual development. Hence, graduate programs provide an education composed of a well-adjusted amount of technical and conceptual classes, with an emphasis on the latter. When selecting their courses, students have to decide wisely and keep in mind the ultimate objective of the MFA program, which is to offer intellectual instructions in digital arts.
3. **External Activities:** Another expectation that Chinese students often have is to be involved in commercial projects during their course of studies in the programs, as it is customary in Chinese Universities. But American graduate programs rarely get involved with commercial projects and students are not expected to be directly involved with the professional market during their graduate studies. Although many students might work on the side on some professional projects, these works are not considered as part of the requirements of the curriculum and don't qualify towards the evaluation of students' progress.
    a. On the other hand, students are encouraged to participate in collaborative projects with outside organizations and individuals, in order to elaborate innovative and interdisciplinary approaches that emphasize creativity and enhance their artistic sensibilities.
4. **Student Involvement:** Chinese students usually expect lecture-based courses, where information is exclusively provided by the professors and students have more of a passive/receptive role. But American classrooms, especially at graduate level, often have a different communication pattern, which requires a high level of participation from students and is more interactive. Many graduate classes are conducted as seminars, where constructive discussions between students and professors are customary. Analysis and critiques of artworks consume a considerable portion of class time, and although professors lead the arguments and direct them in the right paths, eloquent contributions from the parts of students are the norm and anticipated. American students are usually well accustomed with criticism of their own as well as other students' works, since the practice has been integrated in their culture for decades. In his book entitled

The Public and its Problems, John Dewey (1927) had already underlined the importance of having the opportunity for productive discussions and debates, which he characterized as the embodiment of reflective thinking. Although not usually one of their strangleholds, but this is an area in which Chinese students can easily proliferate, if they make the most of their traditional socialization and teamwork skills. According to Richard Nisbett (2004), for Asians "success is often sought as a group goal rather than as a personal badge of merit. For them, feeling good about themselves is likely to be tied to the sense that they are in harmony with the wishes of the group's expectations" (p. 49). It is necessary to highlight the importance of team work for digital artists in general, since due to their technical complexities, digital art projects usually require efforts from the parts of many individuals simultaneously. In such scenarios, being able to work harmoniously in a group is certainly a big plus.

a. Consistent with their adaptive qualities noted earlier, it is conspicuous that the majorities of Chinese students judiciously follow the guidance of their instructors and are able to grasp the importance of self-expression in a timely manner. Taking full advantage of their strong oriental heritage helps them to eventually develop their own artistic trajectories and increasingly attain a distinctive and emphatic presence in the digital arts arena. As noted by Picquart (2011), "Chinese students have an increasingly stronger voice in the world. They are not 'passive beneficiaries' of occidental currents of thoughts anymore. On the contrary, they are ever more inclined to elaborate their own ideologies and ways of life" (p. 162).

5. **Vocational Plans:** Upon entering the programs, Chinese students have specific career goals in mind, which most often are related to the animation, visual effects, and gaming industries. But they soon realize that even though graduates of the MFA programs can follow a variety of career paths, including advertising and entertainment related professions, one of the main goals of the MFA programs is to produce artists and intellectuals who are able to teach in academia.

SOLUTIONS AND RECOMMENDATIONS

Beneficiating from their solid art foundation skills, Chinese students should focus the majority of their time and efforts in their graduate programs to improve their knowledge of contemporary art theories. They would need to keep in mind that their final portfolio should not only demonstrate skillful implementation of technology, but more importantly, reveal strong conceptual growth and attentiveness to scholarly research based on their creative exploration of new media technologies. The style and content of their works should divulge their personality, feelings, ideas, and viewpoints in an educated and sophisticated fashion.

To help students understand that computers and digital media are powerful tools of expression, the programs seek to expose them to a wide array of creative approaches that incorporate critical thinking and the use of advanced multimedia. Past and current digital art forms are examined in detail, in an effort to prepare individuals capable of defining future pathways at the intersection of the arts, sciences, and new technologies.

Since digital arts are closely related to previous art movements, it is imperative for students to expand their knowledge of Western art history and philosophy of art, in order to define their work in the context of a historical continuation. Consequently, it is recommended that during their first year of studies, they enroll in various art history and art theory classes. This will be beneficial to them in the long term, and they will be able to apply the acquired knowledge to their art practices industriously. Furthermore, it is advisable that students who plan to continue their digital arts education in the US prepare themselves accordingly in advance and gain fundament knowledge of Western art history and theory prior to their relocation.

Research is an integral part of the creative practice and students are frequently asked to perform thorough conceptual and technical investigations on the subject of their choice before starting to work on their hands-on projects. Each project necessitates its own specific investigative approach. In other words, research is a project based process that differs every time based on each individual case. As an example, assuming that a student project comprises of building an app that translates words into sounds and colors, research on the relationships between sound, emotions, and brainwaves, as well as color theory and user accessibility might become indispensable. At times, technical research might also become necessary, since in some cases the technology for the execution of a project might not be readily available. In those situations, students might need to spend a significant amount of time and effort exploring different options to find and sometimes invent the appropriate technological solutions to their needs (Candy, 2011, p. 8).

Due to the complex nature of creating digital artworks, they realizations often require cooperative efforts among multiple individuals. Hence, students are encouraged to participate more actively in interdisciplinary teamwork. They are expected to seek opportunities in joining forces with faculty and peers from related areas such as visual arts, design, computer science, and other academic areas from within the university. Chinese students should rely on their strong heritage of social collaboration to fulfill this requirement effectively.

FUTURE RESEARCH DIRECTIONS

US graduate schools should recognize that Chinese students' needs might be slightly different than those of their American peers. A cross-cultural recognition between professors and students is necessary. During the first year, professors should consider granting students more extensive timeslots to develop their projects' ideas, and offer them widespread guidance to help them shift from a mainly technical background towards a concept-oriented training. Courses in contemporary art theory as well as art history and philosophy of art should be recommended to them from an early stage. In the classroom, they should be frequently encouraged to voice their opinions regarding their studied topics and dynamically share educational experiences with other classmates.

Perhaps, a suitable future research opportunity would be to investigate the different ways that American institutions could create a more productive academic setting for Chinese students as soon as they enter the United States. Consequently, this growing population of students will be prepared for the requirements of their majors as early on as possible and will avoid wasting a significant amount of time to discover these requirements the hard way, on their own.

CONCLUSION

It is clear that digital arts will continue to evolve with countless possibilities in many different grounds. It is also gratifying to notice that the related programs in the US have attracted so many talented students from China. Most of these students have proven to have strong work ethics and determination. Provided with supportive, demanding, and constructive educational environments, they certainly can prevail not only in their studies but also in their consecutive professional careers.

REFERENCES

Bachelard, B. (1942). *Water and Dreams: An Essay on Imagination of Matter* (E. Salsali, Trans.). Paris, France: Corti.

Buci-Glucksman, C. (2001). *The Esthetic of Time in Japan: From Zen to the Virtual* (E. Salsali, Trans.). Paris, France: Galilée.

Cain, S. (2012). *The Power of Introverts in a World that Can't Stop Talking*. New York, NY: Crown.

Candy, L., & Edmounds, E. (2005). *Interacting: Art, Research and Creative Practitioner*. Faringdon, United Kingdom: Libri.

College Art Association. (2015). MFA Standards: Statement on Terminal Degrees in the Visual Arts and Design. *College Art Association.* Retrieved February 18, 2016, from http://www.collegeart.org/guidelines/mfa

Dewey, J. (1927). *The Public and its Problems*. Athens, OH: Ohio University Press.

Misco, T. (2015). *Preparing to Succeed at US Colleges and Universities: A Guidebook for Chinese Students.* Self-published. Printed by CreateSpace Independent Publishing Platform.

Nisbett, R. (2003). *The Geography of Thought: How Asians and Westerners Think Differently ... and Why*. New York, NY: Free Press.

Paul, C. A. (2005). *Digital Art*. New York, NY: Thames & Hudson.

Picquart, P. (2011). *China in Twenty Years, and the Rest of the World* (E. Salsali, Trans.). Lausanne, Switzerland: Favre.

Xueqin, J. (2011, July 19). Thinking Right: Coaching a Wave of Chinese Students for College in America. *The Chronicle of Higher Education.* Retrieved March 16, 2016, from http://chronicle.com/article/Thinking-Right-Coaching-a/128287/

Zhang, L., & Sternberg, R. (2012). *Culture and Intellectual Styles*. New York, NY: Springer.

Zuo, M. (2015, May 29). 8000 Chinese students in US expelled for low grades, academic dishonesty, breaking rules. *South China Morning Post*. Retrieved March 22, 2016, from http://www.scmp.com/news/china/money-wealth/article/1812115/roughly-3pc-chinese-students-us-expelled-low-grades-academic

KEY TERMS AND DEFINITIONS

Dada: A European art movement from the early 20th century. Public participation was one of the important characteristics of this art form.

Fluxus: A European art movement from the late 50s.It mainly rejected the standards and conventions of traditional arts in favor of new forms, which at times were interactive and cross-disciplinary.

Interdisciplinarity: Combining multiple disciplines, technics, or media together in order to achieve new methods of practice and thought in different fields.

Physical Computing: Refers to interactive computer systems that can sense real-world physical activities and react to them accordingly.

Real-Time Interactivity: Refers to applications in which computer's feedback is available virtually immediately to the user. In general, 3D interactive applications need capable video cards and high processor power to respond in real-time to the operator's input.

Sonic Art: An artistic discipline in which sound is used as a medium and can take various digital and interactive pathways.

Visual Effects: The processes in which computer imagery and animation are integrated into live footage to create the final pictorial illusion for the viewer.

Chapter 6
Crowdfunding, Entrepreneurship, and Design Education:
Case Study of a Campaign by Design Students from China

Zhou Ding
Nanjing University of the Arts, China

Jiang Jiabei
Nanjing University of the Arts, China

ABSTRACT

The purpose of this paper is to develop further insights into micro-entrepreneurship programs participated in by Chinese industrial designers. A model of creative thinking is employed to explain the campaign creation process. A case study research in sample design entrepreneurs was designed and conducted, and it was composed of three steps: preparing for data collection, collecting the evidences, and analyzing the evidences. It was found that five main defects in creative thinking work as obstacles to crowdfunding success. In order to overcome these drawbacks, it is suggested that designers involved in micro-entrepreneurship programs should acquire the abilities of building prototypes, following the design procedure, finding and solving problems, defining ideation and applying the evaluation methods. Current findings and future study can contribute to the curriculum development for China's industrial design education.

DOI: 10.4018/978-1-5225-0911-0.ch006

INTRODUCTION

The Chinese government is making heavy investments to convert grass-roots innovation into business. In Shanghai, Communist Youth League supported the 2012 Shanghai Maker Carnival. The construction of more than 100 government-supported innovation houses is proposed by local officials. Some hackers' fabrication laboratories in China earn official supports which take the form of latest facilities or considerable allowances usually used as the rents (Parker, 2014). Around the globe, online crowdfunding currently becomes a significantly effective approach to realize a wide range of designers' micro-entrepreneurship dreams. A number of campaigns have been successfully launched on Kickstarter that one of the most popular international crowdfunding platforms. On the community of Kickstarter, a product design campaign for a smart watch named Pebble attracted more than $10.3 million from 68,928 backers as of May 2012 (Newman, 2012).

In the field of design education, what can domestic educators in China learn from these changing external conditions? Few researchers in China have so far studied this emerging question. Focusing on university-level design education and industrial business, two scholars Rodber and Wormald (2007) in the UK pointed out that the process of creating wealth through new-product development has changed since the early 21st century. Similar to Chinese government, British government strongly supports science and innovation to create innovative goods and services for economic success. Design thinking, marketing-oriented new product development as well as the integration of design system and service are highlighted as the important drivers. Thus two researchers suggested orienting industrial design education (IDE) to the latest trends so as to alleviate the mismatch between IDE curricula and newly required skills and knowledge in industries. While this suggestion can serve as a reference for this paper, it is of limited helps for China's IDE and crowdfunding campaign creation.

Despite the rapid growth of online crowdfunding, there is comparatively little research on fund-raising campaigns. Focusing on the participants' psychology and behaviour, a study proposed enhancing the motivations of both the creators and the supporters via crowdfunding support tools (Hui, Greenberg & Gerber, 2013). Gerber and Hui (2013) suggested applying the principles of connecting participants by, for example, creating a reward system and inspiring social feedbacks (Sakamoto et al., 2013). Kuppuswamy and Bayus (2014) indicated that project creators should pay more attention to the first and last periods of fundraising campaigns, in which platform backers tend to contribute more than in middle weeks. However, these prior researches regarding campaign management may rarely benefit entrepreneurs in their campaign creation.

In order to research online crowdfunding, a case study was conducted here targeting a group of industrial design students (hereafter referred to as the Students) at Nanjing University of the Arts, China. The Students created a smart ring for which the fundraising campaign was unsuccessful on Indiegogo, a large global crowdfunding and fundraising online site. Through figuring out how the Students carried out product design campaigns and why their creative thinking-based campaign creation failed to reach the funding goal, this research aims to develop further insights into the design micro-entrepreneurship based on China's IDE.

After presenting the key concepts and theoretical basis, this paper gives a systematic description of the research process, which is divided into three main sections. The first section details the process of preparing for data collection, including case selection and the formulation of the evidence-collecting procedure. The second section describes the collected evidences. Through analyzing the evidences, Section three concludes that the Students' crowdfunding failure was the result of their creative thinking shortcomings, including deficient knowledge about prototype construction, poorly designed procedure, poorly defined target problem and solution, ambiguous function definition, and disorderly evaluation process. The research findings herein can potentially benefit not only micro-entrepreneurs engaged in crowdfunding design but also local design educators in China, who significantly need to rethink about IDE and develop related curricula to satisfy the changing industry's needs.

KEY CONCEPTS AND THEORETICAL BASIS

Crowdfunding and AON Model

Crowdfunding, defined as the integrated practice of raising funds from a large number of people particularly over the Internet (Drake, 2014), has gained rapid development since the beginning of this century. Following ArtistShare, the first online crowdfunding platform set up in 2003, related sites have mushroomed around the globe since 2005. More than one million individual campaigns were carried out globally in 2012, and the industry size would expectedly grow to US$5.1 billion in 2013 (Barnett, 2013).

Indiegogo, the research setting of this study, is one of the largest crowdfunding platforms worldwide. Its funding model is referred to as All-or-Nothing (AON), the most widespread and effective crowdfunding type (Weisul 2014). Based on it, project founders follow three main steps to realize their entrepreneurship projects. The initial stage is to start a complete campaign for a designed product. Then the product is released on Indiegogo platform. Meanwhile, crowdfunding work, including campaign material preparation, project marketing and project goal following (Hui et al., 2013), is implemented to raise funds. Following the above two steps, if the total funds meet the expected minimum before the expiration date, the project founders will definitely gain all the funds and rewards for future project implementation; otherwise, the project has to be sadly terminated by Indiegogo, and all the invested capital will be directly returned to investors. This AON model is typical and even imitated by numbers of crowdfunding platforms in China, like Demohour and Zhongchou. Therefore, case of the Students on Indiegogo has general significance in Chinese micro-entrepreneurship and industrial design context.

Creative Thinking

This research was carried out in the context of new product development, a complete process of bringing a new product to market. Authors tended to evaluate the effective transformation of market opportunities into products available (Krishnan & Ulrich, 2001) for sale as 'creative'. And the science behind the phenomenon is the designed products satisfy customer needs and wants. Runco and Chand (1995) defined creative thinking as a mental process of activity leading to original and adaptive ideas, solutions, or insights. Product design, as a kind of creative behaviour, is typically defined in terms of originality and adaptiveness that together follow the integrated processes generating the above results. In addition, two scholars presented a two-tier model of creative thinking in detail.

As shown in Figure 1, knowledge and motivation are two basic elements of the first tier. In general, knowledge is declarative or procedural. Declarative knowledge represents fact-based information, while procedural knowledge indicates specialized knowledge which usually requires explicit instructions and strategies. Motivations are classified into two types including intrinsic motivation and extrinsic motivation. Furthermore, the second tier of creative thinking consists of problem discovery, ideation and evaluation, wherein problem discovery implies identification and definition of problems; ideation refers to the ideational process which is described as divergent thinking or associative thinking, while ideational flexibility, namely the variety of ideas, is one of the most important factors for divergent thinking; evaluation indicates various evaluative processes which could influence creative thinking. The model below was employed as a theoretical basis in this paper because it could bring a systematic framework for the following evidence collection. This framework demonstrates an understanding of creativity theories and concepts related to the topic of designers' micro-entrepreneurship and the broader areas of

Figure 1. The model of creative thinking includes two components on the first tier while three ones on the second
Source: Runco and Chand, 1995

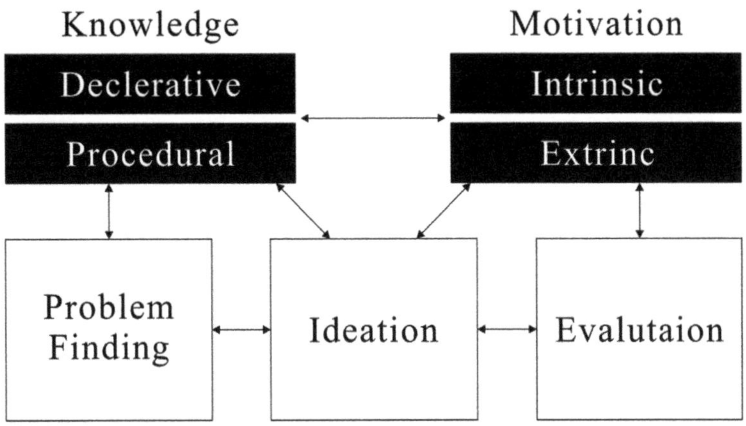

knowledge considered. Selection of the two-tier model of creative thinking depends on its appropriateness, explanatory power, and convenience in application.

In general, the theoretical basis of creative thinking strengthens current researches in four crucial ways (Anonymous, 2016). Firstly, an explicit statement of theoretical assumptions of creative thinking enables readers to critically evaluate the evidence collection of selected case. Secondly, this theoretical basis associates researchers of this study with existing knowledge, especially with regard to practical creation. Guided by the two-tier model of creative thinking, authors are provided with basic selection of research methods. Thirdly, articulation of the theoretical assumptions of current researches requires researchers to solve the problems about how product design campaigns are carried out by designers and why their crowd funding campaigns end up with success or failure. It enables researchers to complete intellectual transition from general description of an observed phenomenon to generalization of various aspects of the phenomenon. Last but not least, application of theories helps authors of this study identify the limitation to the generalization. A theoretical framework specifies essential reasons which influence a phenomenon and highlights demand for examining diversification of those reasons and conditions for the diversification. Therefore, by its application, the two-tier model of creative thinking is valuable because it fulfills the primary purpose of current researches, so that authors of this research may apply that knowledge and understanding to perform more effectively.

RESEARCH DESIGN

A research design is the document of the study, which is created to seek answers for the researched questions. The design of this study effectively defined the aim, objective, scope, method, and data collection approach of the research. Based on critical evaluation of China's IDE in the current context of micro-entrepreneurship, the research aims to develop further insights into the design micro-entrepreneurship based on China's IDE. The research aims to work out two questions, namely, how the product design campaigns are carried out by designers and why their crowd funding campaigns end up with success or failure. Scope of the research is narrowed to designers' campaign creation on a crowd funding platform in view that crowd funding has become a new opportunity for designers' entrepreneurships in China and even the whole world.

As a type of micro-entrepreneurships of emerging designers, the number of crowd funding campaigns launched by Chinese designers is still highly insufficient. Therefore, an inductive approach based on qualitative case research was adopted (Eisenhardt, 1989; Yin, 2013). Bell (1999) also pointed out that "a case study approach is particularly applicable to individual researchers because it realizes certain research of one aspect of a problem within a limited time scale". In order to explore the solutions for research questions which possibly benefit design entrepreneurs and educators in China, the evidences shall be collected through interviews and observations, rather than experiments. As a result, case study became the most suitable method for this kind of researches (Yin, 2013).

Furthermore, the single-case holistic design is deemed as one of the most effective types of case study design, as this research aims to realize objective understanding of the environments and conditions, wherein crowd funding campaigns are created. The conclusion of current researches towards a typical object should be helpful for deepening the comprehension of similar cases. The technological issues of the holistic single-case study mainly focus on design of an effective investigation and interview for data collection, so it is divided into three parts: preparation for data collection, evidence collection, and evidence analysis. The first part consists of case selection and formulation of the evidence collection procedure; based on the model of creative thinking, an investigation framework is formulated and applied in this part. In the second part, the collected evidences are presented systematically. In the third part, the answers to the research questions will be proposed based on the evidence analysis.

PREPARING FOR DATA COLLECTION

Case Selection

When selecting a subject for a case study, researchers used information-oriented sampling here, which is opposed to random sampling. A representative and typical sample meets selection standards of a holistic single-case study, wherein qualitative evidence indicates the knowledge of a current research question. Therefore, a case of Chinese designers' micro-entrepreneurship became an ideal research object.

Authors of this research selected a product campaign on Indiegogo crowdfunding platform, which was launched by Chinese designers from domestic IDE system. The story of an entrepreneurial team, which was a group of industrial design students in China led by Youlin Chen, Zhengpeng Wu and Shuwen Xiao, was selected as the study case. This selection was made based on the following facts. First and foremost, the three group-mates were senior students at the School of Industrial Design, Nanjing University of the Arts (NUA) in China. NUA also is known as the Nanjing Arts Institute, is a university offering undergraduate and graduate degrees in fine arts, design, and related subjects. The institutional history begins in 1912, with the foundation of the Shanghai Chinese Art College. The name was changed to the Shanghai Academy of Fine Arts of in 1930. The amalgamation of several universities in 1952 produced East China Arts College, situated in Wuxi. In 1958, the East China Arts College relocated to Nanjing, and in 1959 it acquired its present name (Anonymous, 2014). So far, the Ministry of Education and China Industrial Design Association still have a crucial influence on whole China's IDE system, in which most schools share the similar curricula and standards in design education. Therefore, the Students at NUA were rather typical representatives. Second, the entrepreneurial team had converted creative thinking into entrepreneurial practice by launching a crowdfunding campaign on Indiegogo, thus it was positively feasible to closely observe their campaign creation for collecting targeted evidences. Their campaign project, Rilove Smart Ring, is a gesture-based wearable device being able to control music, camera, phone calls and even home appliances. Supported by 68 investors investing a total of US$1,935 as of the expiration date, this campaign finally failed to reach its US$20,000 goal and was therefore terminated. Last but not least, the authors of this research taught at NUA, the Students' university, and had obtained the Students' consents to the investigation and interview.

Formulation of the Evidence-Collecting Procedure

In order to collect systematic evidences, this task was performed in six steps based on the model of creative thinking: declarative knowledge, procedural knowledge, motivation, problem finding, ideation and evaluation. As shown in Table 1, a framework, in which the aim of investigation and the interview questions for each step are defined, is developed:

- **Step for Declarative Knowledge:** At the first step, declarative knowledge including valued and neglected knowledge gained or missed by Students in

Table 1. The case interview framework

	Aim of Investigation	Interview Questions
Declarative Knowledge	Describe valued and neglected knowledge in China's IDE system.	• Did you learn any kind of design knowledge at the university which contributed to the campaign creation? If yes, please describe it/them. •Did you learn any kind of design knowledge outside the university which contributed to the campaign creation? If yes, please describe it/them. • Was there any kind of design knowledge which you intended to apply but failed to, or about which you had to consult others? If yes, please describe it/them.
Procedural Knowledge	Describe the Students' procedural knowledge.	• Was there an explicit instruction that guided your campaign creation process? If yes, please describe it/them. • Please describe your campaign creation process. In other words, how did you create your campaign? • Was there any strategy applied in your campaign creation process? If yes, please describe it/them.
Motivation	Describe the intrinsic motivation and/or extrinsic motivation of the Students.	Please describe your motivation for the campaign creation.
Problem Finding	Describe the Students' problem-finding process.	• What was the starting point of your campaign creation? • How did you identify and define a problem?
Ideation	Describe the Students' divergent ideational process and/or associative ideation process.	• Please describe the ideational process of your campaign creation. • How many ideas were generated in your divergent ideation process—if you experienced such a process?
Evaluation	Describe the Students' evaluation process.	• Was there any evaluation process for your campaign creation? If yes, please describe it/them and answer the following question. • Was there any modification after the evaluation? If yes, please describe it/them.

current China's IDE system was surveyed. The valued knowledge stands for elements of fact-based information such as 3D modelling software, product engineering, and material applications, which could effectively promote the campaign creation. The neglected knowledge presents a series of elements in design process, wherein the Students intended to apply them but failed; or they had to consult others for these elements.

- **Step for Procedural Knowledge:** At the second step, procedural knowledge of the Students was investigated mainly based on their specialized experience. Procedural knowledge instructs them to organize campaign creation and apply creativity strategies during design.
- **Step for Motivation:** In order to interview the Students' possible intrinsic motivation and extrinsic motivation, the third step was organized in this framework.
- **Step for Problem Finding:** At the fourth step, the possible problem-finding process of the Students was surveyed. Does problem discovery only exist when a problem works as a starting point of campaign creation? To some extent, the quality of problem discovery depends on how a creator identifies and defines a problem.
- **Step for Ideation:** The fifth step aims to describe the Students' ideational process of campaign creation, which may be divergent or associative. In a possible divergent ideational process, the quantity of generated ideas is taken as one of the most important influential factors.
- **Step for Evaluation:** Step six aims to investigate a possible evaluation process of the Students, wherein the modification might be set up to improve the campaign creation.

COLLECTING THE EVIDENCES

Through summarizing the results of the interview with the Students, the detailed case evidences are listed in Table 2:

- **Evidences of Declarative Knowledge:** In terms of declarative knowledge supporting campaign creation, the Students declared that they learned digital modelling skill, products styles definitions, and ergonomics principle at school, and referring to online examples for product style and building visual communication outside school. However, current China's IDE provides deficient information with regard to engineering structures, prototypes building, interactive hardware or software.

Table 2. Case interview evidences

	Detailed Items	Evidences
Declarative Knowledge	Design knowledge learned at school	• CAD skills, especially in digital modelling. • Definitions of product styles. • Ergonomics.
	Design knowledge learned outside school	• Designing a stylish product through referring to online examples. • Visualizing a campaign for communication through referring to online examples.
	Design knowledge that is deficient	• Designing product engineering structures. • Building product models or prototypes. • Developing hardware or software related to interactive design.
Procedural Knowledge	Explicit instruction	N/A
	Campaign creation process	• Designing product function, form and business model. • Evaluating design through investigation. • Modifying design. • Building prototype.
	Applied strategy in the process	• Brain storming. • Online investigation template.
Motivation	Intrinsic motivation	• Satisfying personal interests. • Pursuing a sense of accomplishment. • Exploring future career opportunity.
	Extrinsic motivation	• Investment from partner. • Possible economic benefits.
Problem Finding	Starting point of campaign creation	• Defining a ring combining technology and fashion.
	Definition of problems	N/A.
Ideation	Ideational process	• Generating product form. • Altering product function until confirmation.
	Divergent ideation process	Generating eight divergent ideations for product style.
Evaluation	Evaluation process	• Self-evaluation. • evaluating product form and function by conducting tests on a small number of users.
	Modification process	Modifying product form and function through self-evaluation and testing.

- **Evidences of Procedural Knowledge:** According to the Students' description, explicit instruction was not included in their procedural knowledge. In order to create a product campaign, the Students scheduled and practised a flexible creation process: designing multiple elements of product, evaluating product design according to investigation, modifying final outcome, building full function prototype. Also, the Students applied strategy in the process, including brain storming and template-based online investigation.
- **Evidences of Motivation:** Basically, interview information showed that the Students worked with clear intrinsic motivation, such as satisfying personal interests, pursuing a sense of accomplishment, and exploring future career opportunity. Because of investment from partner and possible economic benefits, the Students' campaign creation also was driven by extrinsic motivation.
- **Evidences of Problem Finding:** The Students started the campaign creation from defining a ring combining technology and fashion, but not a problem that needs to be solved. Namely, in the process of creativity, the Students ignored problem finding that an essential component of creative thinking system.
- **Evidences of Ideation:** In order to create a high quality campaign, the Students carried out an ideational process in which product form was generated and its function was altered until confirmation. This ideational process producing eight diverse ideas for product style is typically divergent.
- **Evidences of Evaluation:** The Students applied a series of evaluation process, first of all, self-evaluation. In addition, conducting tests on a small number of users, the Students evaluated product form and function. According to self-evaluation and testing, the Students furthermore modified product form and function.

ANALYZING THE EVIDENCES

In order to evaluate the Students' creative thinking, the above evidences were analyzed in the following six sections:

1. **Declarative Knowledge Analysis:** The Students' declarative knowledge was concentrated in the field of design aesthetics. And online resources also provided access to its development. However, a lack of practical knowledge resulted in the Students' failure to build a truly applicable prototype.

2. **Procedural Knowledge Analysis:** Basically, the Students received no explicit instructions. Despite the application of some strategies, a poorly designed procedure still complicated their campaign creation.
3. **Motivation Analysis:** The Students owned clear intrinsic and extrinsic motivations.
4. **Problem Finding Analysis:** The Students didn't build the campaign based on problem finding. Consequently, this campaign lacked adequate attraction to the target participants, because specific problems were not solved.
5. **Ideation Analysis:** The Students struggled to generate a complete ideation because of ambiguous function definition. Divergent thinking rather than associative thinking was applied in their ideational process.
6. **Evaluation Analysis:** The Students consciously evaluated the campaign, but disorderly evaluation process might obscure the guideline for modification.

In summary, the above analysis illustrates how the Students carried out product design campaigns. Another research question, why the Students failed to carry out a successful campaign, also is effectively explained. There are five main defects in the Students' current creative thinking: deficient knowledge about prototype construction, poorly designed procedure, poorly defined target problem and solution, ambiguous ideational definition and disorderly evaluation process, all of which acted as the obstacles to their crowdfunding success. Associative thinking was additionally evaluated to assist with the ideational process. Overcoming these drawbacks requires reasonably developing creative thinking.

LIMITATIONS AND FUTURE RESEARCH DIRECTIONS

The first limitation lies in the research scope: this research only targets crowdfunding, and other types of designers' micro-entrepreneurships in China are not involved. The second limitation is that the research design in this paper is made for single sample. Yin (2013) defined this type of design as single-case holistic design and also presented the other three more complex types of case study design: single-case embedded design, multi-case holistic design and multi-case embedded design.

In terms of future research, crowdfunding will still be the focus of our research, considering that this type of micro-entrepreneurship is highly accessible to Chinese designers. In addition, an integrated research will be designed to gain a deeper comprehension of the relationship between crowdfunding success and creative thinking elements. Multiple cases, particularly successful product-design campaigns, will be included in future research. Last but not least, both current and future findings will be employed to develop university-level curricula for China's IDE. Experimental courses, which aim to enhance the creative thinking of designers, will expectedly be carried out and evaluated at NUA.

CONCLUSION

In the context of China's advocate of designers' micro-entrepreneurship, especially AON crowdfunding, domestic IDE and prior research can hardly produce ideal solutions to the emerging problems. How to effectively carry out a product design campaign? Which factors determine the success or failure of a fund-raising campaign? In order to answer these questions, a case study on a Chinese students' design team, which launched a crowdfunding campaign, was designed and conducted herein. In this study, evidences of the campaign creation process were collected based on the model of creative thinking. Analysis of the evidences revealed that the failure in the Students' fund-raising campaign could be attributed to creative thinking obstacles: deficient knowledge about prototype construction, poorly designed procedure, poorly defined target problem and solution, ambiguous ideational definition and disorderly evaluation process. This result objectively reflects the main defects of current China's IDE. Therefore it is required that related curricula and courses, which aim at improving the design entrepreneurs' creative thinking, should be positively developed. To be more specific, designers involved in micro-entrepreneurship should possess the strong abilities of building prototypes, following the design procedure, finding and solving problems, defining ideation and applying the evaluation methods. In future research, further insights into the relationship between the crowdfunding result and the elements of creative thinking will be developed, and possible research findings might be transformed into university-level curricula for China's IDE.

ACKNOWLEDGMENT

This article is one result of the research project XJ2014019 at Nanjing University of the Arts, China. During the period that the lead author worked as a visiting scholar in New Zealand, China Scholarship Council and Victoria University of Wellington's School of Design importantly supported his research work. He particularly thanks Dr. Anne Galloway, a senior lecturer in design culture and context at Victoria University of Wellington, for her tutorial of design research method.

REFERENCES

Anonymous. (2014). *About*. Retrieved September 19, 2014, from: http://en.nua.edu.cn/s/92/t/585/p/21/c/1863/list.htm

Anonymous. (2016). *Organizing Your Social Sciences Research Paper: Theoretical Framework*. Retrieved May 11, 2016, from: http://libguides.usc.edu/writingguide/theoreticalframework

Barnett, C. (2013). *Top 10 Crowdfunding Sites For Fundraising*. Retrieved November 11, 2014, from: http://www.forbes.com/sites/chancebarnett/2013/05/08/top-10-crowdfunding-sites-for-fundraising/

Bell, J. (1999). *Doing your research project: a guide for first researchers in education and social science*. Academic Press.

Drake, D. (2014). *Crowdfunding: It's No Longer a Buzzword*. Retrieved November 11, 2014, from: http://www.crowdsourcing.org/editorial/crowdfunding-its-no-longer-a-buzzword/32268

Eisenhardt, K. M. (1989). Building theories from case study research. *Academy of Management Review*, *14*(4), 532–550.

Greeberg, M. D., Hui, J., & Gerber, E. (2013). Crowdfunding: A resource exchange perspective. In *Proceedings of CHI '13 Extended Abstracts on Human Factors in Computing Systems*.

Hui, J., Greenberg, M., & Gerber, E. (2013). Understanding crowdfunding work: implications for support tools. In *Proceedings of CHI '13 Extended Abstracts on Human Factors in Computing Systems*.

Krishnan, V., & Ulrich, K. T. (2001). Product development decisions: A review of the literature. *Management Science*, *47*(1), 1–21. doi:10.1287/mnsc.47.1.1.10668

Kuppuswamy, V., & Bayus, B. L. (2014). *Crowdfunding creative ideas: The dynamics of project backers in Kickstarter*. UNC Kenan-Flagler Research Paper, (2013-15).

Newman, J. (2012). *Pebble Smartwatch Pre-Orders Are Sold Out, $10+ Million Pledged*. Retrieved November 12, 2004, from http://techland.time.com/2012/05/10/pebble-smartwatch-pre-orders-sold-out/

Parker, E. (2014). *The Chinese Government Is Investing Heavily in the Maker Movement*. Retrieved May 14, 2014, from http://www.slate.com/blogs/future_tense/2014/05/14/the_chinese_government_is_investing_heavily_in_the_maker_movement.html

Rodber, M., & Wormald, P. (2007). Aligning industrial design education to emerging trends in professional practice and industry. In *DS 43: Proceedings of E&PDE 2007, the 9th International Conference on Engineering and Product Design Education.* University of Northumbria.

Runco, M. A., & Chand, I. (1995). Cognition and creativity. *Educational Psychology Review*, *7*(3), 243–267. doi:10.1007/BF02213373

Sakamoto, M., Nakajima, T., Liu, Y., & Todorka, A. (2013). Design and evaluation of micro-crowdfunding: Encourating sustainable behaviour in micro-level crowdfunding. In *AcademicMindTrek '13 Proceedings of International Conference on Making Sense of Converging Media.*

Weisul, K. (2014). *In Crowdfunding, All-or-Nothing Campaigns Are More Successful.* Retrieved November 12, 2014, from http://www.inc.com/kimberly-weisul/why-your-crowdfunding-campaign-should-be-all-or-nothing.html?cid=em01011week25day19d

Yin, R. K. (2013). *Case study research: Design and methods.* Sage Publications.

KEY TERMS AND DEFINITIONS

AON Crowdfunding Model: A type of crowdfunding model in which project founders can gain all money if total funds meet expected minimum on expiration date, otherwise their campaigns must be terminated and all invested capital will be returned to investors. It is believed that AON fundraising campaigns involve substantially larger capital goals, and are much more likely to be successful at achieving their goals, because AON projects typically provide more detailed information on the campaign.

Campaign Creation: A process of creating a project for raising funds. A campaign is any series of actions or events that are meant to achieve a particular result, like an advertising campaign of television commercials and Internet ads that tries to convince kids to buy bubble gum-flavored toothpaste.

Commercial Wealth: Abundance of valuable resources or valuable material possessions in commerce.

Creative Thinking: A mental process of activity leading to original and adaptive ideas, solutions, or insights. Creativity is a phenomenon whereby something new and somehow valuable is formed. The created item may be intangible or a physical object. Creative thinking is a way of looking at problems or situations from a fresh perspective that suggests unorthodox solutions. Creative thinking can be stimulated both by an unstructured process such as brainstorming, and by a structured process such as lateral thinking.

Crowdfunding: A form of alternative finance, which has emerged outside of the traditional financial system. The practice of funding a project or venture by raising monetary contributions from a large number of people, today often performed via internet-mediated registries, but the concept can also be executed through mail-order subscriptions, benefit events, and other methods.

Industrial Design Education: A process of design applied to products being manufactured through multiple techniques. Industrial design is a process of design applied to products that are to be manufactured through techniques of mass production.

Micro-Enterprise: A small business generally employing 10 people or less, and have a capital asset of less than PhP 3,000,000. Internationally, most microenterprises are family businesses employing one or two persons.

Micro-Entrepreneurship: A process of starting a small scale business or other organization. The basis of micro-enterprise in the U.S. is micro-entrepreneurship, recognizing a fundamental right of people to apply their individual talents, creativity and hard work to better their lives.

Chapter 7
Industry:
Academia Partnerships in Design Education

Jui-Che Tu
National Yunlin University of Science and Technology, Taiwan

Yu-Chen Huang
National University of Kaohsiung, Taiwan

ABSTRACT

"Industry-academia partnership" refers to the cooperative relationships between universities and enterprises during economic development. The purpose of this study is to establish a cooperative channel, strengthen the two-way exchanges between universities and enterprises, and highlight their complementary relationship. Universities shall guide enterprises toward correct operational philosophy, and assist industrial upgrading and economic development, while enhancing the research level of universities. As a result of this study, industry-academia partnership integrates social resources to give full play to mutual exchange between economic functions and social adaptations in order to promote industrial upgrading.

DOI: 10.4018/978-1-5225-0911-0.ch007

INTRODUCTION

Basically, the goal of university education is to cultivate talents that meet the needs of nations and provide new blood to leading industrial and scientific R&D. However, are talents trained by universities truly consistent with the requirements of a knowledge economy era in terms of expertise and capacity? Furthermore, does the mode of universities leading scientific R&D advance with the times? If not, from the perspective of effective management, the establishment of an appropriate "platform of industry-academia partnership" would be the solution.

Industry-academia partnership has become a vital and expanding topic in Taiwan. Especially, since Taiwan passed the "Fundamental Science and Technology Act" in 2011, which clearly defined the transfer of ownership of intellectual property rights and the outcome from the National Science Council (NSC) to research institutions or enterprises, meaning more cooperative programs have been implemented between universities and industry. If we extend the meaning of the First article of the University Act, universities must maintain basic research, and consider priority and practical research fields in order to support the development of national competitiveness and meet the demands of the rapid development of new technologies. Therefore, the research results of universities shall be converted into new technologies and practical products. If a win-win solution can be found between universities and industry to coordinate the advancement of teaching, research, and production, the development, transformation, dissemination, and application of knowledge and information can be improved in order to drive the rapid development of technology and economy.

In recent years, developed countries have paid attention to industry-academia cooperation, stressed the enhancement of research application results, and driven the growth of industry and national competitiveness via the cooperation of enterprises and universities (Hsiao, 2000). In response to international and regional economic integration, the agenda of new economic and trade orders, such as the WTO, and the impact brought about by newly emerged industrialized countries, enterprises shall strive to improve their technologies and R&D capabilities in order accelerate industrial upgrading(Chen, 2001).

Today Taiwan's economy develops towards the goal of globalization. Taiwan goes through economic reform from traditional economy to new economy, which affects the change and innovation of design education in Taiwan. In early phase, Taiwanese government provided basis for industrial development through state-owned enterprises. In the era of original equipment manufacturer (OEM), Taiwan engaged mainly in pure OEM without product design. Design and formula was provided by commissioned factories. Manufacturers produced products to meet the needs and requirements enterprise customers, which were sold with the trademark of the enterprise customers. (Reference: Economic Globalization. Chapter One:

The history and current status of industrial development in Taiwan. Study on the correlation between public investment and the development of manufacturing and services industries).

Later, along with the accumulation of experience of manufactures, the change of industrial supply chain, and market demands, besides OEM, Taiwanese industry began to have its own design capability and unique formula of products. Original design manufactures (ODM) came into being. Manufactures should have design, R&D, production, and assembly capabilities of products, negotiate product specifications together with ODM buyers, and carry out product design or improvement work.

After the 1990s, with the advancement in technology and design capacity, Taiwanese industry gradually began to have the awareness of Original Brand Manufactures (OBM), followed by the rise of discussion on enterprise transition of "OEM or OBM" by Taiwanese enterprise management science circle (Chen & Lin, 1990; Li & Chen, 1998; Huang, 2002; Hung, 2004; Huang, 2005; Yang, 2005; Li, 2006; Hsu, Chou, and Shih, 2006; Lin, 2006; Chen, 2006). Enterprises began to engage in OBM to build their own brands and exceed the scope of OEM to become the owners of brands. They had marketing channels to sell their products, develop and design their own corporate images, and obtain maximum market economic benefits.

Furthermore, in the face of globalization, development of technology and Internet, maturity of industrial technology, Taiwan ushered in the mode of integrated device manufacturer (IDM). IDM refers to the enterprises which can complete all the works including design, manufacturing, packaging, testing, sales of their own brands IC with high-tech and capital-intensive characteristics. They obtained market share through their technical and financial advantages to form the pattern of oligopoly (Hsu & Jhang, 2008).

In the face of changing industrial environments, only constant research and innovation to develop new products that meet the needs of the market can facilitate enterprises achieving competitive advantages in market competition (Dai, 2000). However, as average SMEs have insufficient understanding of new markets and technologies, and enjoy only a 5% success rate of new product development, they tend to hold a conservative attitude towards the development of new products (Li, 1992). Overall, the quality of enterprise researchers is critical to the enterprises' future; however, in terms of research organization and training, SMEs do not have the matured institutionalization or the foresight of large enterprises, thus, they urgently require assistance with R&D and the establishment of research organizations. In the face of this dilemma, universities can assist enterprises to engage in research and innovation, develop new products, and reserve talents for research and innovation.

On the other hand, through cooperation with enterprises, teachers can convey the latest industry information to students as they accumulate practical experience, which not only improves the quality of teaching, but also ensures that students can adapt to the skills required in the workplace, and become compliant with the standards of professional competency (Bottge & Osterman, 1998). Hence, strengthening the understanding of the status quo and demand of professional knowledge by teachers and students is beneficial for students to successfully adapt to the ever-changing industrial ecology of the future (Lin & Davis, 1997).

In summary, if, during their cooperation, universities and enterprises can adhere to the principles of resource complementation and benefit-sharing, diversified partnership channels can be established between industries and academic circles. Consequently, enterprises can gain more competitiveness in the global market, and the design education of Taiwan, as well as the overall design profession, can be improved to facilitate the industry-academia partnership achieving a win-win solution (Ruth, 1996; Meng & Wu, 2000). This is the main purpose of this study.

Fields and Concept of Industry: Academia Partnerships in Design Education

According to the design correlation, trends of talent nurturing, and course directions, relevant fields for partnerships between industries and universities are selected. The positive influence of enterprises, including technological developments and industrial trends, will be introduced in the teaching resources and curriculum contents of various universities. Meanwhile, it is necessary to establish a database of potential enterprises for cooperation. Teachers and students shall learn the process of design, the R&D of enterprises, and relevant technologies in order to improve their practical abilities (Figure 1).

Regarding the significance of industry-academia partnerships, Bloedon and Stoke (1994) proposed that funding should be undertaken by enterprises for industry-academia partnerships to initiate research activities and plans between higher education institutions and industries. Kang (1997) assumed that industry-academia

Figure 1. Introduction of partnership with enterprises in design education

cooperation should be implemented under the guidance of organizational goals. Universities, and their related social institutions, shall work together comprehensively under the principle of reciprocity to facilitate efficient use of resources or initiate effective integration of theory and practice to achieve goals. Hsiao (1997) also pointed out that industry-academia partnership could allow teachers and students to participate in the work of enterprises for payment, develop products, improve systems, and strive for co-development. Jiang (2000) further asserted that industry-academia partnerships refer to the strategy of cooperation, which integrates teaching resources and industrial technology resources, and combines theoretical teaching and practical application. Huang (2002) also believed that industry-academia partnerships strategically provide students with opportunities to apply theories to practice, and enterprises with talent resources. Wu and Lin (2004) deemed that industry-academia partnerships refer to the cooperation between enterprises and universities. On the one hand, it implements the research of frontier and practical technology universities; on the other hand, it encouraged enterprises to participate in the application research of universities, cultivate potential talents for themselves to achieve the functions of research, training, information communication, and talent cultivation.

To sum up, "industry-academia partnership" denotes the cooperation between enterprises representing economy and a university representing academic research, combines social resources to give full play to two-way exchanges between economic function and social adaption, and improves production technology and industrial upgrading. Its implementation strengthens the two-way communications and relationships between universities and industries. In other words, universities can assist industrial upgrading and economic development, while enhancing education and research, in order to finally achieve regional scientific and technological resource integration (Li, 2010).

Regarding the features of industry-academia partnerships in design education, enterprises can find relevant talents through cooperative activities, and save expenses associated with device support; while universities can fully utilize human and material resources to improve economic benefits. Meanwhile, teachers that participate in cooperative research can convey their teaching experience by combining theory with practice in order to achieve more satisfying teaching and learning results. In short, industry-academia partnerships can improve the technologies of industry, the research level of universities, and design abilities of students.

When planning university design education, there are various training stages to achieve the strategic development of their partner enterprises. First, industry elites will be invited to teach short-term design courses to strengthen students' ability in theory and practice. Second, students will be provided with internship opportunities during the mid-term to directly participate in design development and technologies in the workplace in order to deepen their understanding of workplace environments and strengthen their independent design ability. Third, the long-term goal is to provide employment opportunities to students and match excellent students with relevant enterprises. Finally, such partnerships will promote research with the required characteristics and applications (Chung, 2015).

In order to realize the above mentioned strategic development of industry-academia partnership in the cultivation and training of design education and meet industry-academia demands, reform course design, diverse and innovative thinking patterns have been introduced so that design methods began to paying attention to integrated design course planning with design thinking and creative development.

Keeley et al. proposed "The ten types of innovation" in 1999 which was a framework facilitating enterprises to thinking about innovation systematically and change their viewpoints on innovation from "product" to "multiple innovation" (Kumar, 2009). The framework suggested enterprises to innovate from ten aspects, namely, creative process, core process, product/service performance, service system, customer service, service delivery channel, brand, customer stunning, business model, and value network, as shown in Figure 3.

Figure 2. Strategic development of industry-academia partnership

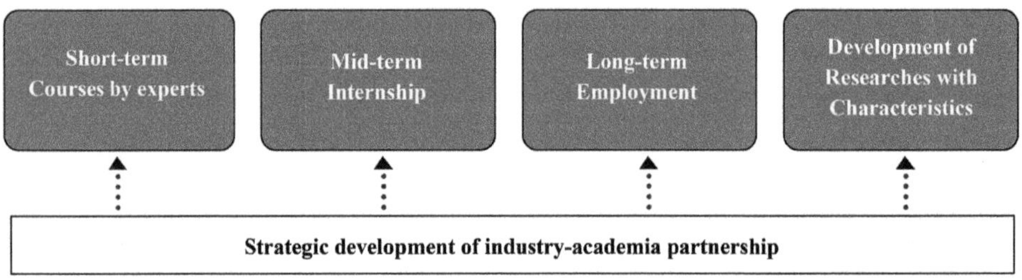

Figure 3. The ten types of innovation
(Keeley et al., 1999)

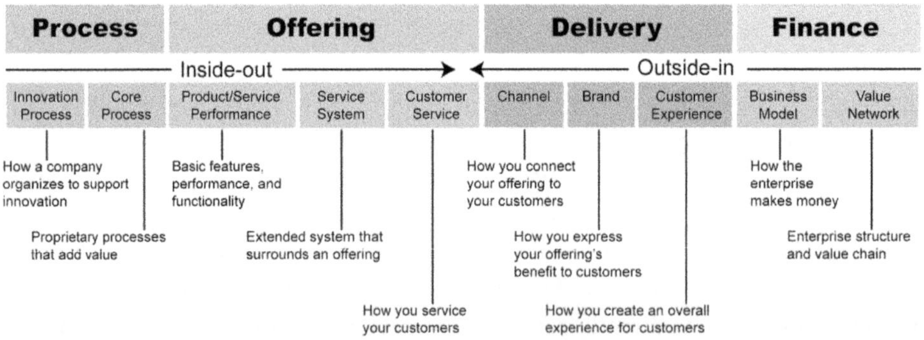

In addition, Design Thinking has been applied in foreign countries for ages. It was proposed by Tim Brown (2008), the President of IDEO, a design company. Tim defined that "Design Thinking is people-oriented design spirit and method, taking people's demands, action, and the feasibility of technologies or business into consideration."Design Thinking is a methodology about innovation and design, helping various topics find innovative solutions and create more possibilities. The concept of Design Thinking represents the conversion from the surface to depth of design, from appearance to the insights of customers' demands and psychology, and from the attention on objects to the thinking of human. In recent years, increasing studies explored the impact of Design Thinking on design education. For instance, Hsieh (2012) took advantage of production waste of enterprises for re-design to strengthen the awareness of environmental protection, analyzed the design thinking capacity of enterprises and the constituent elements of team creativity, and their relationship. He summarized that there were six factors constituting the characteristics of design thinking: namely, integrity, innovation, attitude, and so on. The factor of integrity was the most influential. The characteristics of team creativity included four factors: namely, goal-orientation, communication, integration, and resource sharing and innovation resources. He had found in his studies that design thinking and team creativity were highly relevant and verified that the characteristics of design thinking could have positive impact on the performance of team creativity. In addition, Huang (2012), based on service design, combined concept map to form a mode via which students could conduct design thinking and a system for continuous practice and design. Students could make innovative products meeting industrial requirements while learning. Also, they could learn the design elements of cultural and creative industry. Jen (2014) explored an innovation process centering on user experience. Design Thinking should explore problems and demands from the observation of the connotation of user, facilitate enterprises to enhance the value of their products and services via innovation, and bring about more friendly and convenient life experience. Besides, there were applied researches such as concept generation from the perspective of user experience (Hsu, 2005) and the application of Chinese ancient Thirty-Six Stratagems to generate creative ideas, expand search space, enhance design imagination, interaction, and interest, and stimulate more new ideas (Liu, 1998).

Relationship of Industry: Academia Partnership in Design Education

Kang (1995) mentioned that, sustainable industry-academia partnerships should benefit both sides. Thus, from the standpoints of enterprises and universities, this article discusses the separate and combined benefits of both sides in order to promote the establishment of such cooperation. On the one hand, from the standpoint of enterprises, through cooperation with universities, they can gain professional guidance, the latest and innovative knowledge, and utilize universities' resources (e.g., advanced equipment) to establish technical standards (Peters & Fusfeld, 1982). In addition, enterprise can reduce the R&D risks, time, and costs related to new technologies and new products, solve problems, improve productivity, and stimulate their internal R&D creativity (Geisler & Rubenstin, 1989). Moreover, they can find excellent students to employ after graduation (Liu, 1983), boost their corporate image, and contribute to society (Gao & He, 1996).

From the standpoint of universities, such partnerships allow teachers to better understand the status quo of industry and professional demands, and compile cases and materials for their teaching notes during the partnership (Tu, 1996). Furthermore, it provides both teachers and students with exposure opportunities of actual practices, and offers funds and employment to students (Geisler & Rubenstein, 1989; Tseng, 1990; Dai, 2000).

Specifically, industry-academia partnerships allow for mutual exchange between universities and enterprises, as well as full play of resources, devices, and technologies owned by both sides in order to achieve effective integration of theory and practice. Therefore, both sides are advised to recognize the real benefits of cooperation to achieve win-win goals (Kang, 1995). See the separate and combined benefits of both sides during cooperation in Table 1.

By considering the characteristics of design education, and through appropriate industry-academia partnerships, education, internships, and the teaching and learning effects of design education can truly be enhanced. Meanwhile, more development space can be offered to enterprises, universities, teachers, and students, allowing theory and practice to become closer than ever before. The characteristics of industry-academia partnership in design education are listed, as follows (See Figure 4):

1. Students learn how to cooperate with others during practice related to their design courses, fully understand the objectives of design education, expand their knowledge and skills to learn, and improve their academic performance.
2. From learning design to the application of skills, balanced development between practical experience and professional design knowledge and theories can help students understand the truth of the concepts of design, and the real purposes of design education. Moreover, they will obtain relevant work experience, laying a foundation for their future work.
3. The implementation of industry-academia partnership aims at the problems and demands of research, product development, technical services, talent training, studies of special topics, and industrial diagnosis, provides a platform of mutual

Table 1. Benefit for industry-academia partnerships

Benefits of Industry-Academia Partnerships	Details
Shared benefits	Effective application of the resources of both sides
	Integration of theory and practice
	Achievement of win-win and reciprocal goals
	Enhancement of mutual exchange and cooperation
Benefits for enterprise	Introduction of appropriate development and design concepts
	Gain frontier technologies
	Gain innovative knowledge
	Answers to specific questions
	High-quality talents
	Reduction of development risk and cost
	Stimulation of internal creativity
	Enhancement of corporate image
Benefits for universities	Enhancement of teaching and learning effects
	Financial support
	Access to market information
	Improved research levels
	Enhancement of academic reputation
	Increased employment opportunities
	Promotion of practical experience of teachers and students

(Source: Geisler & Rubenstein, 1989; Kang, 1995; Tu, 1996; Peters & Fusfeld, 1982; Tzeng, 1990; Liu, 1983; Gao & He, 1996; Dai, 2000)

Figure 4. Industry-academia partnership

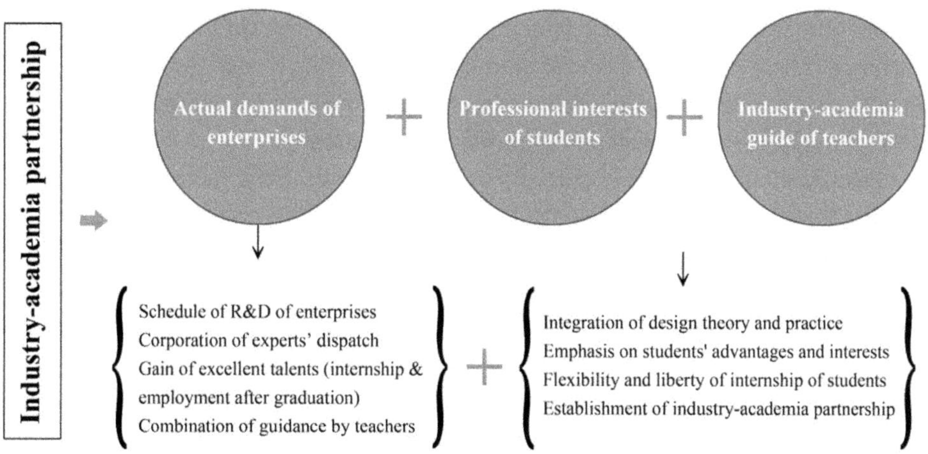

communication between enterprises and universities, promotes integration, and complements cooperation between them by adding practice to teaching.

4. The research results of teachers and students during a partnership benefits both enterprises and society, as it gains ideal teaching materials for design education through the accumulation of cases and summarization of relevant materials. Due to the involvement of vast enterprise experience and methods, design education will become more diversified and practical.
5. Through continuous research, design, and development, professional industry-academia partnerships in design education confirm the combination of design theory and practice, which results in innovative applications that change both sides.
6. Partnerships have positive influence on four parties, students, teachers, enterprises, and society. Students can better grasp the nature of design education through learning the professional knowledge of design education, participation in the practices of enterprises, and combine theory with practice. In addition, the actual interactive and segmented experiments and verification during cooperation between enterprises and universities can refine the majors of universities, and promote development and application.

Design education from the perspective of industry-academia partnership aims to cultivate talents with "pragmatic and practical" abilities and bridge the gap between learning ability and career ability. As of the key elements of "joint cultivation of talents by industry and universities" mean, universities and industrial practitioners shall carry out practices teaching together, adopt the approach of "learning while doing, and doing while learning" through the close industry-academia interaction, help students enhance their career ability and competitiveness. During industry-academia design education, teachers can combine both design theory and practical and industrial technology, make teaching materials more flexible and forward-looking. Meanwhile, teachers can understand the status quo and trend of industry and deeply explore innovative design, brand planning, and marketing design through integrated design courses. In turn, enterprises can obtain novelty and inspiration of product R&D through the innovative design thinking of students.

Specifically, universities, enterprises, and students have their respective and specific functions. Universities shall facilitate "R&D industrialization" and "practical teaching". Enterprises shall utilize talent resources of universities to facilitate "R&D improvement" and "transition and upgrading" and enhance competitiveness. Students shall shorten the gap between knowledge and practice, expose to market functions, and enhance career competitiveness (Zhou, 2013).

Industry-academia partnership in design education aims to obtain win-win and mutually reinforcing results. Enterprises intend to use the human resources of universities with design and R&D energy to innovate their technology or management processes, enhance operational competitiveness and performance, and obtain suitable talents. Teachers can get R&D resources, improve R&D quality and quantity, expose students to practical experience and technology, and increase students' career ability and employment opportunities. Students who participate in industry-academic partnership in design education can establish close and mutual trust relationship through university labs with enterprise owners, companies, and factories. Enterprise owners can provide (even overseas) internships for students to broad students vision and help them enhance their workplace strengths (Huang, 2012).

Generally, in industry-academia partnership in design education, the discussion and analysis of design issues are conducted through various data collection approaches, needs surveys, industry and market research, and expert interviews. Meanwhile, the opinions of enterprise owners are collected for brainstorm and joint discussion. After data collection and referring to enterprise owners' market experiences, the demands and key points of design targets will be finally determined. During the partnership, problems are solved and good design ideas and innovative design are proposed by provoke students' design thinking through teachers' guide on integrated design courses and focused conversation method. Finally, the products designed undergo testing and evaluation in labs. The required outcome of industry-academia partnership and the goals and effect of design education shall be achieved through full communication and interaction between enterprises and universities (Huang, 2012).

The Mode (Procedures) of Industry-Academia Partnership in Design Education

Under the trend of globalization, capital, talent, and technology move quickly across countries. And the emergence of international blue ocean market affects the trend of global industry. In the past, Taiwanese industries had laid a solid economic and technological foundation with the co-efforts of enterprise, government, and universities, exceeded the limits of objective conditions, conformed to the trend of globalization, established an industrial network with multinational division of tasks, greatly expanded economic landscape, and created a stunning economic miracle.

Taiwan has entered into agreements like free economic and trade agreement and marched from OEM to OBM. The open political, economic, and educational environment allows the mode of various industry-academic partnerships. Compared with the simple design service of industry-academia partnership in design education (mainly focused on the development of the appearance or structure of products), only the proposal of integrated design and innovation mode of industry-academia partnership in design education can truly meet the demands of industry now. However, innovation model is more than purely design services. It shall provide holistic design services, including product planning and marketing, market research, product positioning and design, packaging design and exhibition design. Various resources like enterprise internships and industry-academia partnership in projects are integrated in teaching to implement the new and innovative industry-academia cooperation model.

In order to implement innovative industry-academia cooperation model, the range of industry-academia partnership in design education includes development, productivity, service, design, academia, and promotion. Under the principles of complementary and reciprocal, in a top-down promotion mode, a partnership shall be implemented in steps, such as planning, execution, control, assessment, etc. The specific mode of the overall implementation includes:

1. **Principles of Implementation:**
   a. To promote the development of industry-academia partnership, universities and educational administrative units shall establish responsible units (such as industry-academia partnership center, collaboration promotion office, technical cooperation office, or design service center), which are responsible for various tasks, such as contacts, planning, design, promotion, and coordination.
   b. The characteristics of universities and the nature and demands of the industrial environment of various regions may serve as the main reference for industry-academia partnership programs.
   c. Regarding the plan or service programs of industry-academia partnerships, enterprises or institutions shall submit an application, which shall be assessed by the responsible units of universities. After discussion, it will be reported to educational administration for verification and transaction.
   d. Regarding the implementation of industry-academia partnership, both sides shall sign an agreement as the basis of their duties, rights, and obligations, which shall record cooperation mode, expected benefits, cooperation projects, effective period, participants, procedures and schedule, materials, equipment and teachers, funds, intellectual property, publishing of results, and final report.
2. **Procedures of Implementation (See Figure 5):**
   a. Search—Partners

In order to promote the plan or service of a partnership, teachers can contact current and previous enterprises, invite/host practice enhancement lectures, seminars, or briefings in order to keep abreast with the status quo, trends, and demands of

Figure 5. Procedures of the implementation of industry-academia partnership

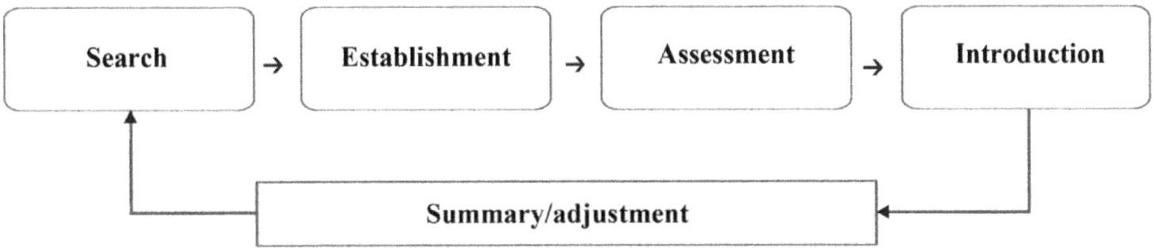

talents, product design, and enterprise development, establish smooth communication channels between universities and enterprises, and reach cooperative agreements.

b. Establishment—Partnership

Through specific lectures, invitation to selection, and visits to enterprises by students, students will deepen their understanding of workplace environments, future industrial developments, connotation of enterprise energies, etc.

c. Assessment--Adaptability of Industry-Academia Partnership
    i. Compliance to the development positioning strength of universities by enterprises
    ii. Professional level of enterprises
    iii. Workplace environments of enterprises
    iv. Devices and other resources of enterprises
    v. Cooperation of professionals at enterprises when teaching at universities
    vi. Future cooperation opportunities with universities
    vii. Feasible internships of students
    viii. Prospect of enterprises
d. Introduction—partners
    i. Signing Memorandum of Understanding, MOU
    ii. Participation in practical courses
    iii. Cooperation during courses taught by professionals
    iv. Offer of internship to students
    v. Research or future education opportunities for teachers
e. Summary/adjustment—modification to partnership

Through an internal work conference on the enhancement of practice of teachers at universities, the directors, teachers in charge, partners of enterprises, and curriculum committees, shall be invited to discuss and adjust practical courses together.

CASES AND EXPERIENCE OF INDUSTRY-ACADEMIA PARTNERSHIP IN DESIGN EDUCATION

Programs of industry-academia partnership in design education started as early as 1964. Wang Yung-chin, President of the Formosa Plastic Group, founded the Ming-Chi Institute of Technology to achieve the educational concept of "learn from practice" through the education mode of "apprenticeship at a factory", and successfully promoted the development of design in Taiwan, which greatly contributed to design education in Taiwan (Li, 1999). During the development of design education, through the establishment and operation of a design center, teachers accumulated experience through practical design work and conveyed it to students via theories. People in charge of design or research programs, in accordance with the actual situation, cooperated with education and allowed students to participate in the implementation of programs, and adhered to the educational mode of "apprenticeship". Students, via study and participation during design practice, grasped both theories and practical skills.

Later, industry-academia partnerships in design education gained more attention. Since 1984, the Industrial Development Bureau, Ministry of Economic Affairs, began to promote design concepts, and through "the program of comprehensive improvement of product design ability", guided SMEs to develop new products, increase product added value, and create profits. According to the Taiwan External Trade Development Council, from 1993 to 2000, 329 cases of guidance were implemented, where concepts related to "diagnostic services for industrial design" and "industry-academia partnership" received good results, which comprehensively enhanced the ability of enterprises in product design and development, and drove the interaction between academia and industry (Chang, 2000). The two cooperative methods are described, as follows:

1. Diagnostic Services for Industrial Design:

Based on the operating status, product planning, structure of the R&D department, design management, and other relevant issues of enterprises, services were provided, including practical diagnosis of industrial design, technical guidance, application of computer technology, green applications, etc. Primarily, teacher's facilitated enterprises to establish product design processes, provided professional design training to researchers of enterprises, and reviewed the defects of product design, and other relevant design and promotional services.

2. Industry-Academia Partnership:

The rich academic resources of Taiwan will be adopted to assist enterprises with potential product development and willingness of development, design products with high added value, improve competitiveness in the global market, and establish an independent design system in order to comprehensively improve their product design ability, especially those with demands for creative products, and thus, give full play to students' rich design innovation.

The status and types of partnership of each country are different. America uses policies and legislation to encourage cooperation between enterprises and universities. Teachers in Japan provide consultancy services to enterprises to improve their product development capabilities. In Germany, enterprises and universities participate together in innovative technology, knowledge diffusion, and development of new products in order to apply theory in practice. Taiwan utilizes academic resources to introduce appropriate design concepts to enterprises and assist their new product development. In reference to the practices of each country regarding industry-academia partnerships, and considering the subjective and objective industrial environments and characteristics of education in Taiwan, different types of industry-academia programs can be developed to meet the actual needs of enterprises and promote the overall development of design education.

Case 1: Application of Stanford Creative Design Thinking in the Development of Brew Ware, Taiwan

1. Case Overview

The import and export trade statistics of Taiwan in recent years have shown that, the output value of industrial products in 2013 and 2014 did not increase much, as influenced by economy and innovative development. Taking brew ware industry as an example, the industry now faces the competition of cheap labor in mainland China, and Europe and America have gradually transferred their orders there. There are many problems in Taiwan's brew ware industry, including the outflow of Taiwanese enterprises (Huidian Market Research Reports Website, 2008). How to gain back markets and competitive advantage is a goal for which Taiwan shall strive. Thus, analysis of this case adopted Stanford creative design thinking, as well as "people-oriented spirit and method of design", as proposed by Tim Brown in Harvard Business Review, in order to "analyze the demands and behavior of humans, and consider feasible designs of science or business." Designers shall observe and simulate the demands and behavior of customers, combine the current professional technology of brew ware, blend innovation, improve the added value of products, enrich product diversity in order to develop products that meet market demands with high innovative value, and improve the competitive edge of Taiwanese industry via product differentiation.

2. Integration of Industry-Academia Partnerships in the Goals of Education

Through observations of current markets and data collection, the case study of industry-academia partnership found that the Taiwanese brew ware industry is facing various challenges, such as technology transfers and fierce competition at home and abroad. Therefore, the plan is to identify key technologies to upgrade the industry, innovative product design, and assist the development of enterprises in order to narrow the gap with foreign counterparts, via product differentiation.

The platform of industry-academia partnership is expected to bring about favorable income, find a creative market niche for the brew ware industry, improve product value and sales and provide solutions to solve real-time problems, actively gain governmental R&D resources, and improve the core competitiveness of enterprises. This case introduced Stanford creative design thinking in the industry. Universities utilized creative ways of thinking to observe the demands of customers, conduct surveys, grasp the development trends of products with commercial value, design creative products, design drafts and make samples, and test product technical usage, which encouraged teachers to blend creative thinking in their education, and closely combined university education with industry. Through interactions between the industry and students, professional, groundbreaking, prospective, and creative design proposals of students have linked with the professional technologies of industry to develop products in a rapid and effective manner to meet the demands of customers. Based on the above cooperation of industries and universities, enterprises have adopted the same amount of cost to generate more profits, thus, improving the added value of products. This outcome met the ideal CP value expected by enterprises, and achieved a win-win situation.

To achieve the above research objectives, this plan sets specific objectives, as follows:

1. Analysis and research related to customers, Stanford creative design thinking, and brew ware.
2. Creative product design and development with Stanford creative design thinking and proposals of creative schemes.

3. Analysis and research related to users.
4. Ideas of creative products.
5. Drafts and samples of creative brew ware.
6. Test of feasibility of creative brew ware.
7. Development of new products with innovation and perceptiveness, offer of important basis for the development of products of Jinhongda Co., Ltd., improvement of added value and sales, and the income of the industry regarding articles of daily use, such as glass containers.

Based on the above proposed objectives, partnerships have helped Jinhongda find other paths for creative products, in addition to the existing professional brewing technologies, which not only improved the impression of existing and potential customers, but also narrowed the gap of the Taiwanese brew ware industry with its counterparts abroad.

3. Characteristics of Design in Industry-Academia Partnerships

The plan mainly utilized Stanford creative design thinking to develop creative products of brew ware, gave full play of the complementation of industry-academia partnerships, enriched product differentiation, and improved the competitive edge of the company. Its design and R&D methods and procedures are, as follows: onsite observations and interviews, analysis of products in the market, establishment of design specifications, designs by Stanford creative design thinking, design of 3D color mode drawing, mode making, qualitative testing, and assessment.

Household utensils change along with improved quality of life. When developing products, current enterprises must consider functionality and emphasize life aesthetics and convenience, which will affect the outcome of products. This study focuses on industry-academia partnership in design education with Jinhongda as the main objectives of the designs of fashionable, convenient, innovative, low-cost, and perspective brew ware. The strengths of the research method include:

1. The application of Stanford creative design thinking and brainstorming by design team members allows designers to effectively understand and meet the demands of users.
2. Grasp new trends and create new value for creative products.
3. Feasibility of technology and profit potential.
4. Visualization and representation of design and development, via skills of the innovative system, for assessment and modification.

5. Presentation of the appearance of products, via computer simulation.
6. Design and research by talents, mutual communication between the industry and universities, and improvement of R&D benefits.

This research focused on the industry-academia partnership with Jinhongda, adopted Stanford creative design thinking, and gained mutual support with professionals via design research with prominent achievements. The product development processes consist of the following five parts: concept generation, mechanical design, appearance design, detail design, product color line, and product sampling.

a. **Concept Generation:** The following 10 concepts came into being via Stanford creative design thinking.
b. **Mechanical Design:** Through the professional assessment of experts in enterprises and design teams, the following three groups of concepts with good functions have been selected for detail design.
c. **Appearance Design:** Through the professional assessment of experts in enterprises and design teams, 8 mechanical designs with good feasibility have been selected for appearance design.

Figure 6. Conceptual plan 1

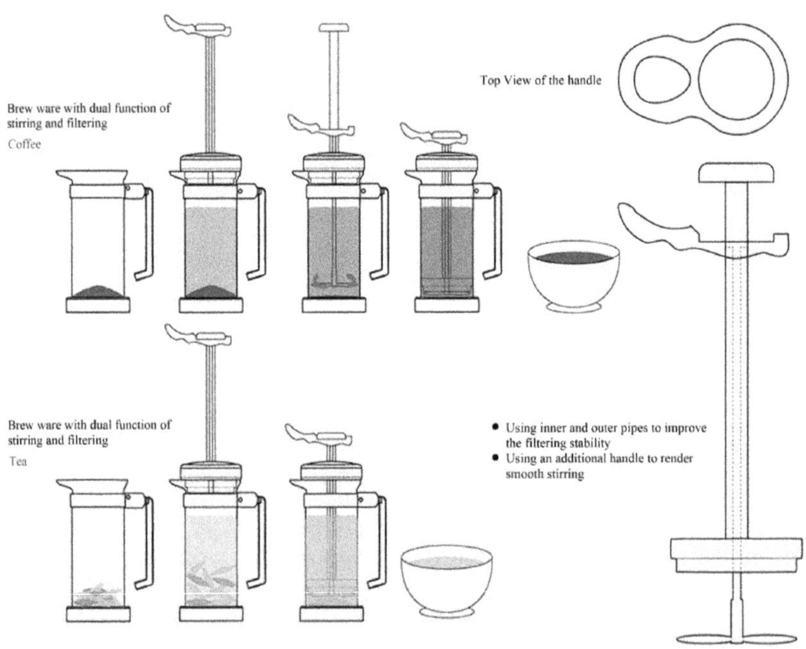

Figure 7. Conceptual plan 2

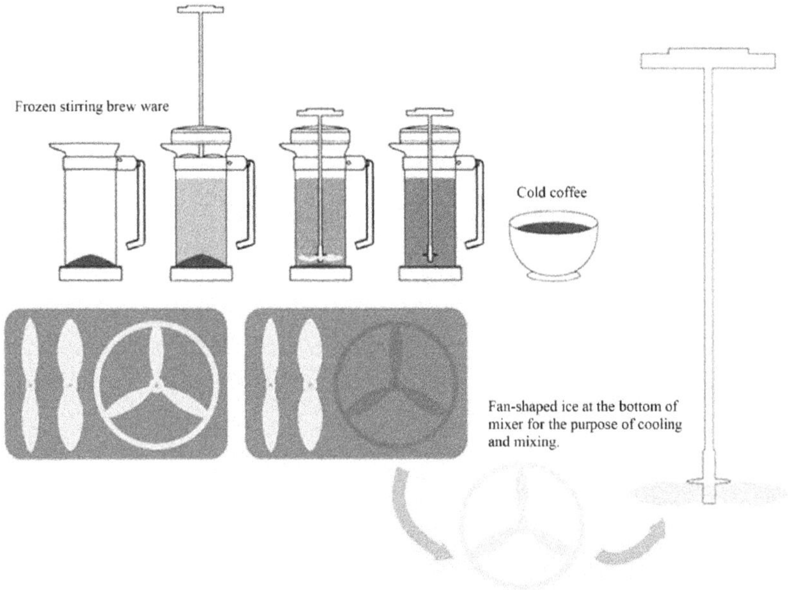

Figure 8. Conceptual plan 3

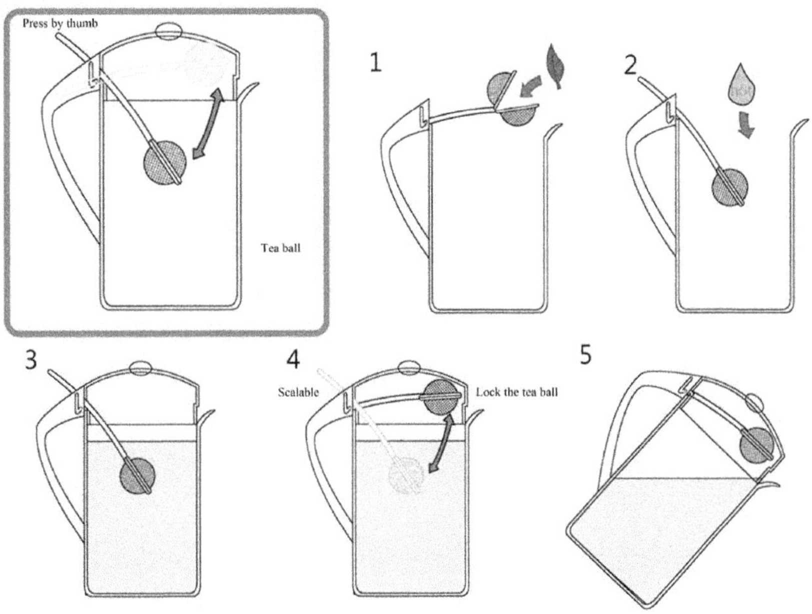

Figure 9. Conceptual plan 4

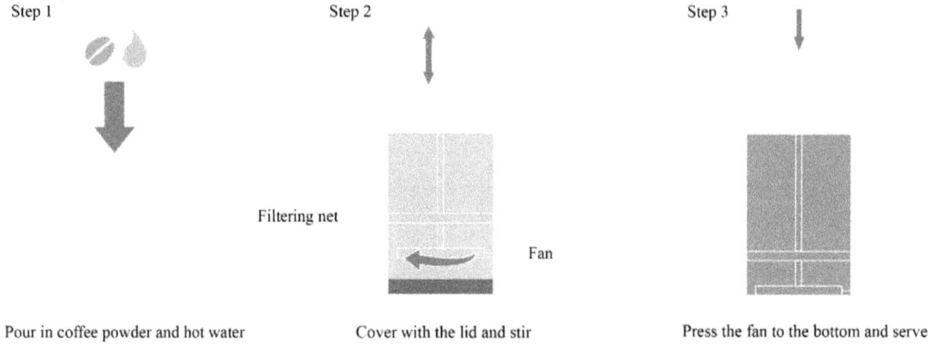

Figure 10. Conceptual plan 5

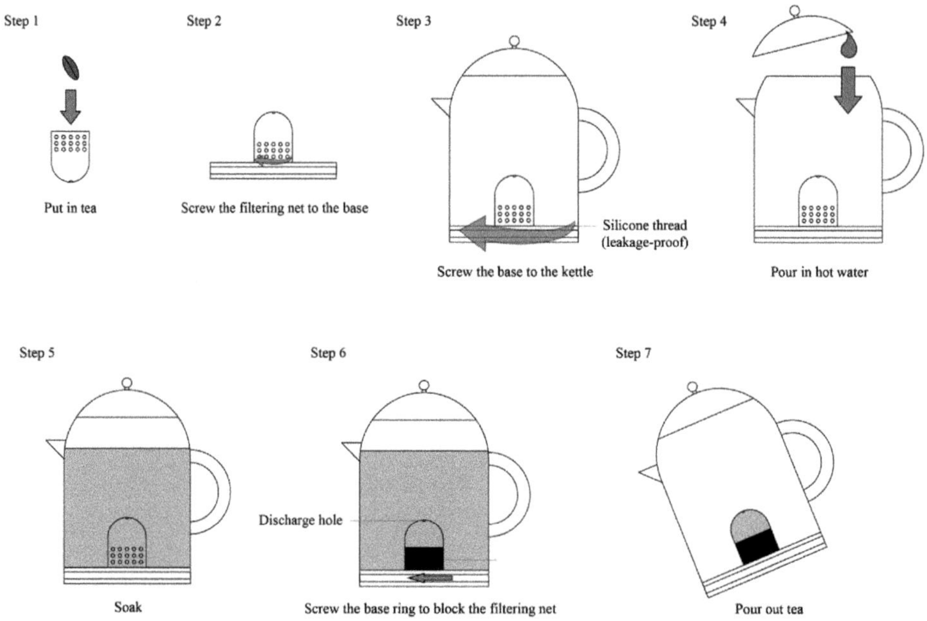

Figure 11. Conceptual plan 6

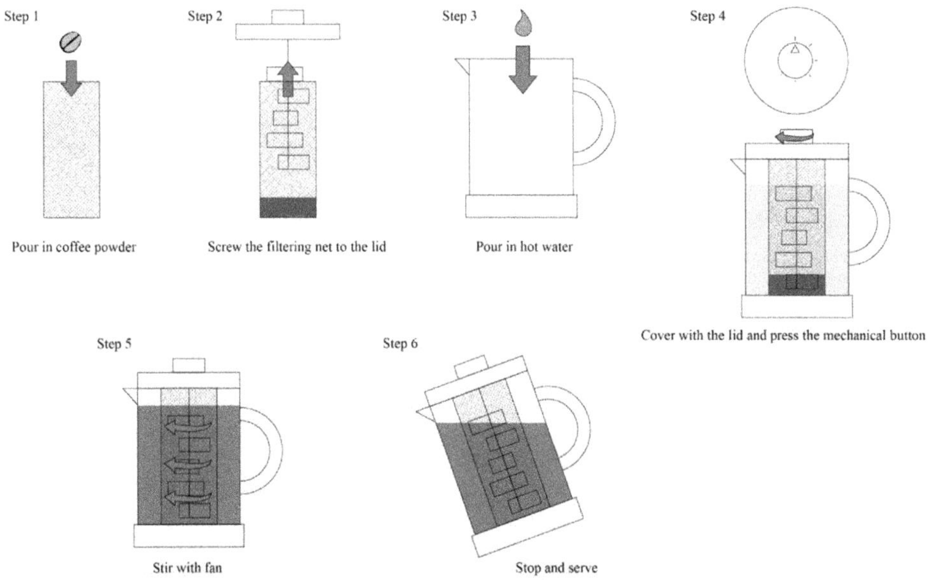

Figure 12. Conceptual plan 7

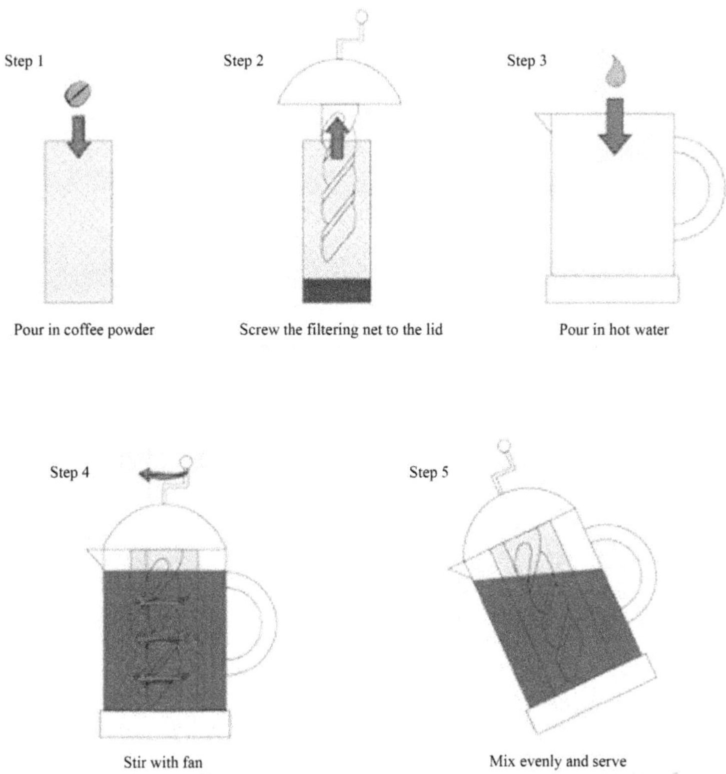

Figure 13. Conceptual plan 8

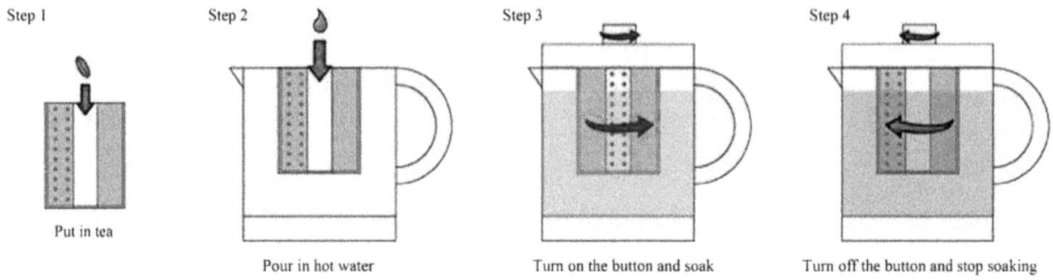

Figure 14. Conceptual plan 9

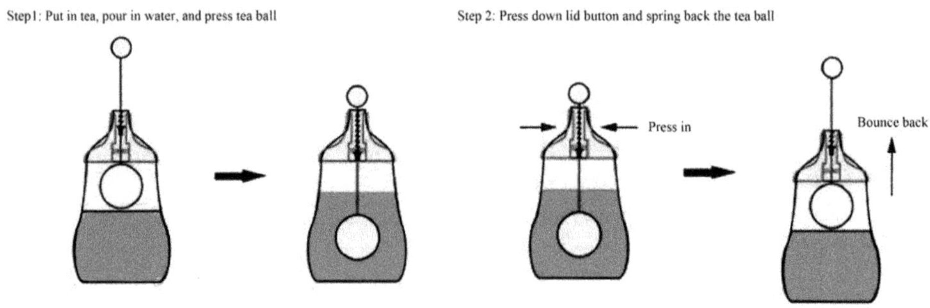

Figure 15. Conceptual plan 10

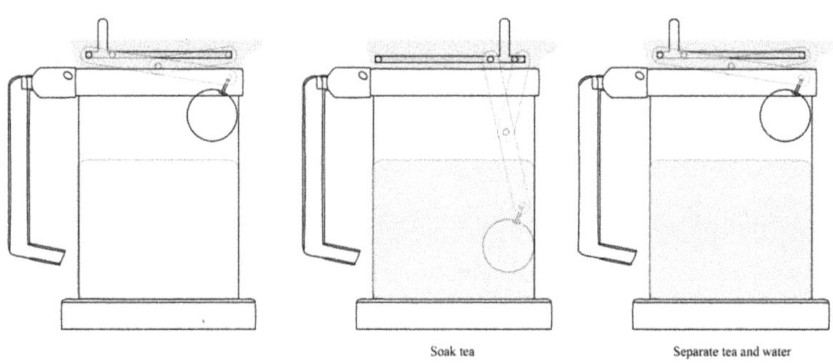

Figure 16. Mechanical design 1

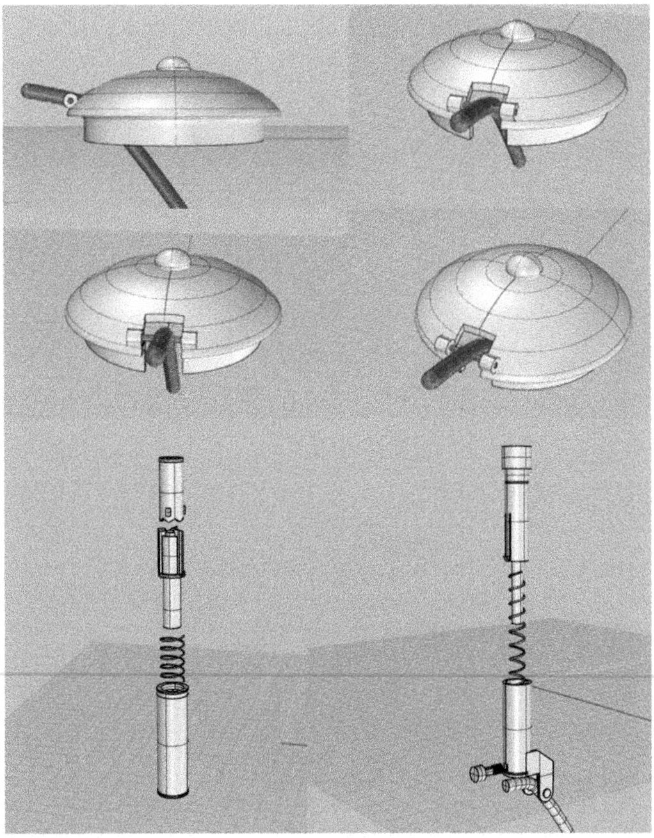

Figure 17. Mechanical design 2

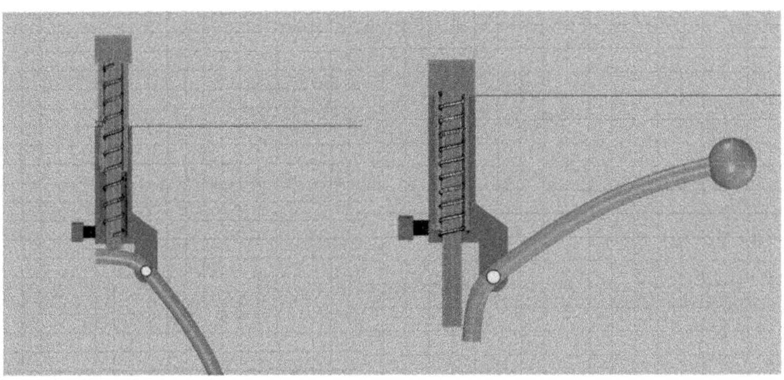

Figure 18. Mechanical design 3

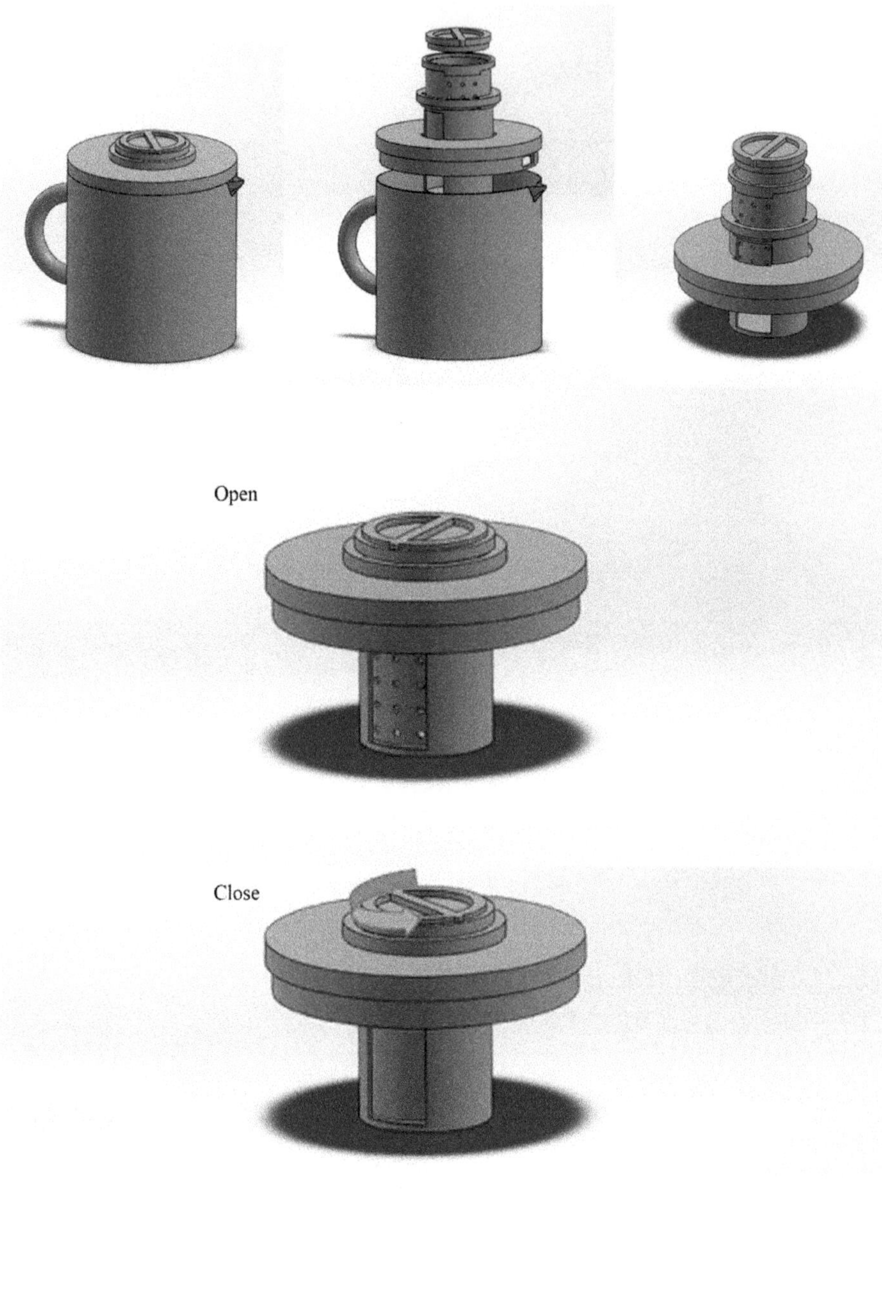

Figure 19. Mechanical design 4

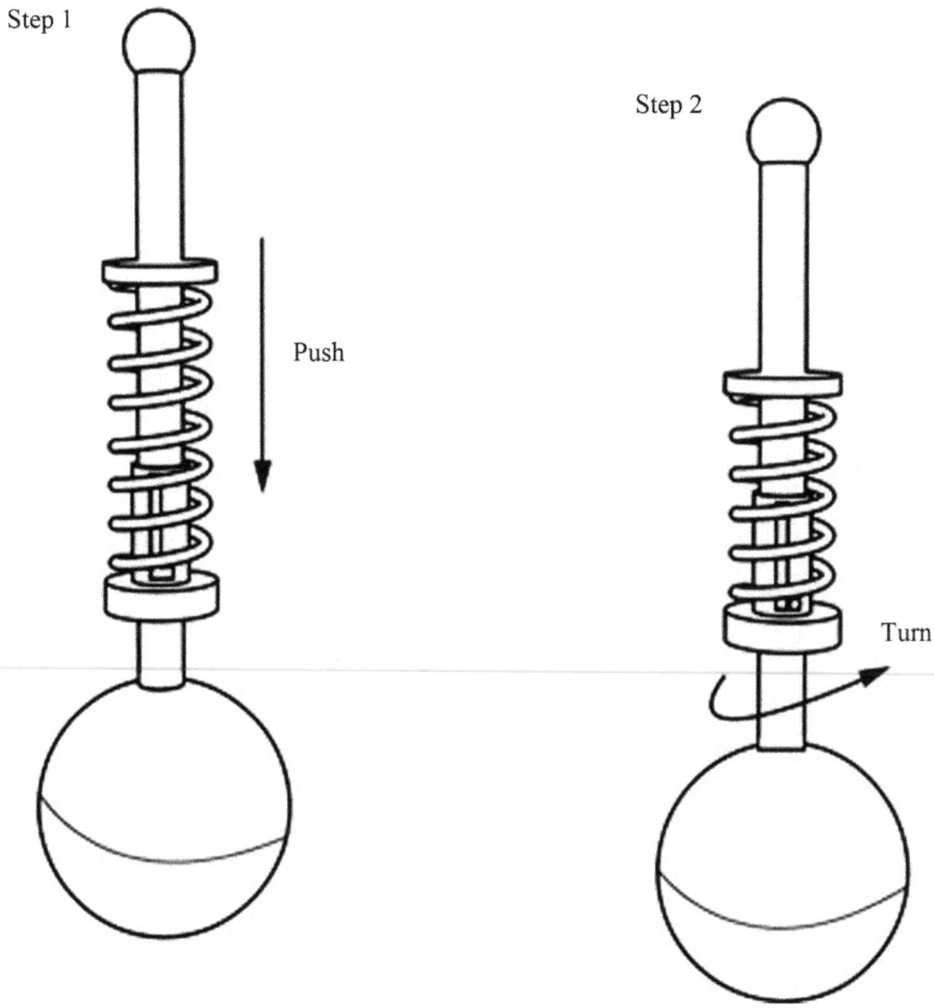

Figure 20. Appearance design 1

Figure 21. Appearance design2

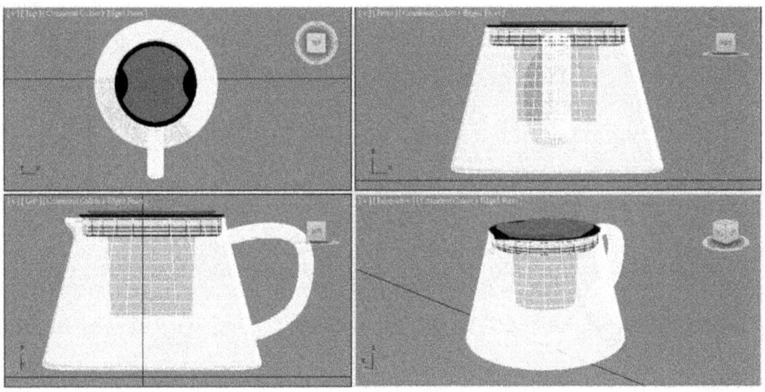

Figure 22. Appearance design 3

Figure 23. Appearance design 4

Figure 24. Appearance design 5

Figure 25. Appearance design 6

Figure 26. Appearance design 7

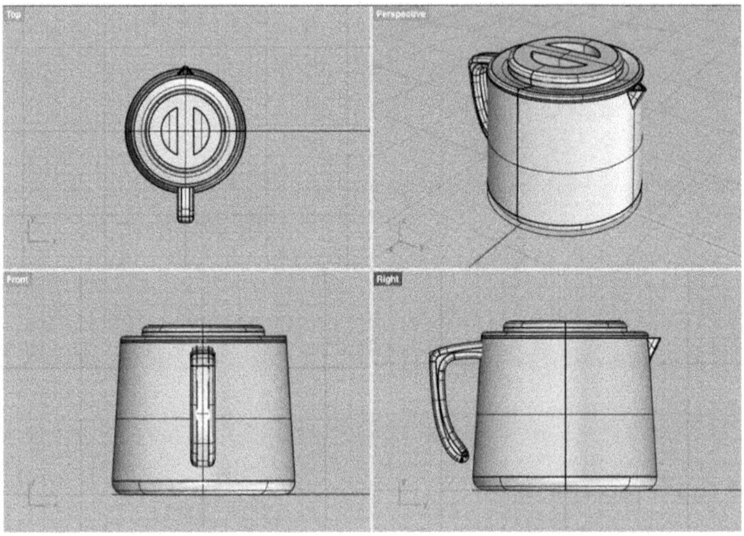

d. **Detail Design:** Through the professional assessment of experts in enterprises and design teams, 6 appearance designs with good feasibility have been selected for detail design.

[Detail 1] Hollow out the lid and utilize the elasticity of plastic to match the lid with the bottle so that when users pour tea, the lid will not fall down.

[Detail 2] The shell and the lid are connected with mortise and tenor, thus, when users turn the inner layer, the shell will not move.

[Detail 3] Connect the upper lid with the lid via spring.

e. **Product Color Line:** Color the final proposal of products to simulate the actual appearance. In accordance with different materials, 2 types of design patterns were proposed.

f. **Product Sampling:** During the development of the product, 3D printing of samples was conducted in order to detect problems in actual operation, while checking the size of each detail to reduce loss and the cost of development.

Case 2: Creative Plan and Design of the Theme Pavilion of Horse's Hoof Clam in Kouhu Township, Taiwan

The government assisted a traditional industry enterprise in crisis to seek a sustainable way forward, and proposed relevant policies of technological upgrading and industrial restructuring, and in 2003, the "plan to promote the manufacturing industry to develop the tourism and leisure industry" came into being. The Executive Yuan

Figure 27. Sectional drawing of detail design 1

Figure 28. Sectional drawing of detail design 2

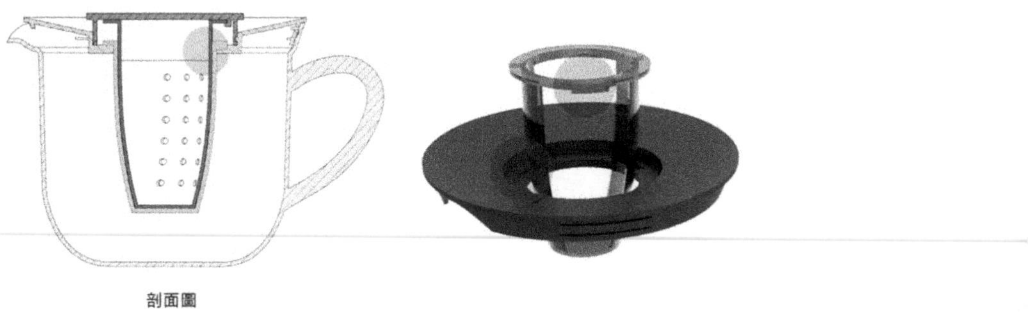

Figure 29. Sectional drawing of detail design 3

included the "development plan of cultural innovation industry" into "Challenge 2008 National Key Development Plans (2002-2007)" in 2002,and announced the governmental intention to develop innovation fields and determine how to combine culture and economy to develop cultural innovation industry and achieve value integrating manufacturing, history, culture, and tourism. The combination of history

Figure 30. [Pattern 1] Plastic for inner layer

Figure 31. [Pattern 2] Stainless steel for inner layer

Figure 32. 3D printing samples of brew ware

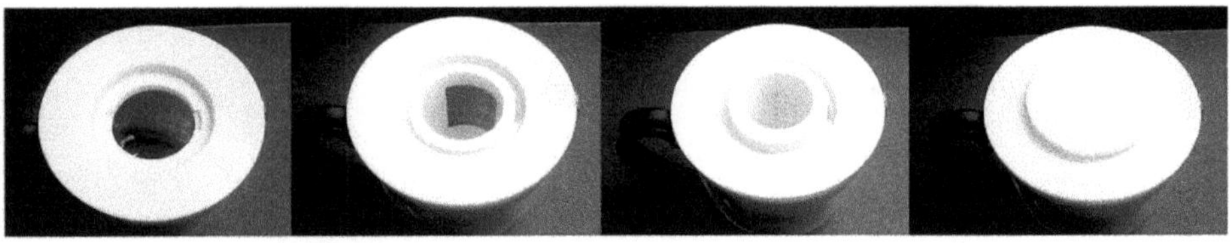

and culture with tourism by the manufacturing industry not only deepen tourism knowledge, it also brought about more business opportunities for survival and development, which extended the development of industrial culture and constantly promoted the sustainability of innovative enterprises.

The plan of an industry-academia partnership with Mei Sin Water Fowl Co., Ltd. promoted the creative plan and design of the Theme Pavilion for Horse's Hoof Clam. Such cooperation among industry, government, and universities developed a tourism industry with high added value and local cultural features, which improved both the environment and quality of life in Taiwan, and promoted the development of a tourism economy both at home and abroad. The plan introduced cultural creative elements in space planning and design, creative product design, brand visual graphic design, etc., which not only promoted the transformation, sales, and added value of traditional aquaculture, but also enhanced the inclusiveness of local culture and facilitated the development of the industry. The plan, under the guidance of Stanford creative design thinking, established appropriate development and application modes of the tourism industry, combined with cultural brands and designs with local characteristics, perceptiveness, and breakthroughs, provided creative cultural products to the "creative plan and design of the Theme Pavilion of Horse's Hoof Clam" of Mei Sin Water Fowl Co., Ltd., rebuilt its brand image, and achieved important results in creative space planning.

1. Case Overview

Kouhu Township, Yunlin County, which takes aquaculture as one of its main industries, is located in the southwest sea of Taiwan. As the traditional culture-based fisheries gradually declined due to financial difficulties and the impact of population outflow, many operators relied on subsidies or financial aid. However, due to different local cultures, terrain, ecological changes, and self-cultivation, its aquaculture was rich in variety, and most local villages participated in the industry. The stable products were oysters and clams.

Recently, Kouhu decided to promote its beautiful fisheries resources and culture, and the Theme Pavilion of Horse's Hoof Clam of Mei Sin Water Fowl Co., Ltd. transformed the simple traditional aquaculture into an entertainment and leisure theme pavilion with cultural and creative characteristics, which deepened people's understanding and recognition of the traditional aquaculture of Kouhu (See Figure 33). Driven by the policies of cultural and creative industries, and the unique cultural

Figure 33. Theme Pavilion of Horse's Hoof Clam of Mei Sin Water Fowl Co., Ltd.

elements of Kouhu, the plan is expected to increase the added value of aquaculture products, enrich product differentiation, create brand charm, and derive cultural and creative products, which can help develop local aquaculture and tourism, and develop products that meet market demands with high innovative value.

2. Integration of Industry-Academia Partnership in the Goals of Education

The plan implemented industry-academia partnership from three aspects, i.e., culture, industry, and tourism (See Figure 33), centered on "intelligence", "creation", and "knowledge", found key technologies, invested design resources to assist industrial upgrade and development, increased the added value, and offered employment opportunities. It further explored how to integrate the aquaculture of Kouhu into a cultural and creative industry, effectively utilized cultural resources, enhanced corporate image, shaped Kouhu culture, and established product brand in order to

Figure 34. Culture, industry, tourism, and Theme Pavilion of Mei Sin Water Fowl Co., Ltd.

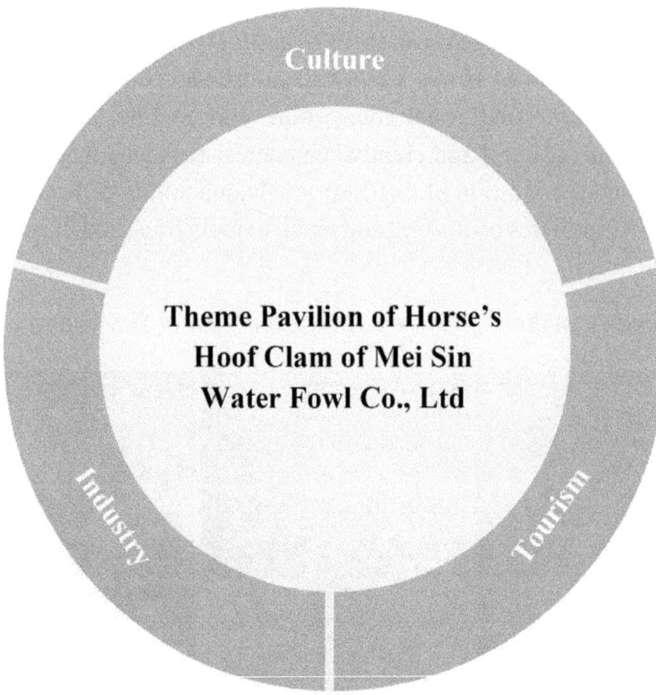

promote local tourism, and improve economic resources and income. See below for the specific plan:

1. Establishment of standard operating procedures for the aquaculture of Horse's Hoof Clam of Taiwan, cultural and creative design and development of products, and design of brand image.
2. Establishment of operating procedures and methods in line with the trend of the aquaculture of Horse's Hoof Clam of Taiwan, cultural and creative design and development of products, 2D illustrator, and 3D sketch up model.
3. Establishment of operating procedures and methods in line with the trend of the aquaculture of Horse's Hoof Clam of Taiwan, cultural and creative design and development of products, and 2D and 3D models.
4. Sampling of the product design and development, and theme model of the aquaculture of Horse's Hoof Clam of Taiwan.
5. Gain assessment results of the aquaculture of Horse's Hoof Clam of Taiwan.
6. Establishment of a complete series of products of the aquaculture of Horse's Hoof Clam of Taiwan, design and development of cultural and creative products, and space design specifications of the spatial design of the theme pavilion.
7. Understanding the actual situation, brand image design, and development process of the aquaculture of Horse's Hoof Clam of Taiwan

The goal of the cooperative plan was to design Horse's Hoof Clam products with unique local characteristics of Taiwan, rebuild brand image, and create products with high inclusiveness via cultural, innovative, and new methods. The plan focused on cultural and creative elements, combined industrial brand image and market demand, and developed a successful tourism industry. It mainly adopted "Stanford design thinking, innovative methods, and procedures", designed tourism with local cultural features, utilized the advantages of industry-academia partnership, enhanced product differentiation, improved competitive edge, and promoted the sustainable development of this creative industry. Its creative design, development methods, and procedures are described, as follows.

Stanford Design Thinking innovative method and procedures:

Step 1: Empathize—in-depth exploration of the problems and demands of users
Step 2: Define—summary of the characteristics and demands of target groups
Step 3: Ideate—imagine optimal solutions and ideas
Step 4: Prototype—producing creative products or services
Step 5: Test—testing the model and degree of satisfaction of demand

3. Characteristics of Design in Industry-Academia Partnerships

This plan conducted an industry-academia partnership with Mei Sin Water Fowl Co., Ltd. by combining the support of enterprise professionals; through the Theme Pavilion of Horse's Hoof Clam, transformed the simple sales methods of Mei Sin to demonstration marketing, built a theme pavilion with cultural and innovative value, promoted people's understanding and interest of folk culture, and brought commercial opportunities to Kouhu Township. The following shows four design proposals, including corporate image shaping (CIS), mascot design, development of a cultural and innovative mall, and field planning and activation, as shown in Figures 35-38:

THE INFLUENCE OF INDUSTRY-ACADEMIA PARTNERSHIPS IN DESIGN EDUCATION ON TEACHERS AND STUDENTS

In the social trend of applying knowledge in practice, diversified intelligence, and development, governmental departments strive to promote industry-university cooperation, establish a model of education based on student employment, innovation, and entrepreneurship, balance theory and practice, and fill the talent gaps of key industries. Through close "industry" and "academia" cooperation, both sides cultivate talent together. Enterprises can participate in the course design of universities, assist teachers' teaching activities, and provide relevant devices for the study of students or opportunities for internship and employment; while teachers, through such cooperation, can update teaching content. Universities can share research equipment and cultivate practical talents via the mode of "practicing while learning and learning while practicing". Design education through industry-academia partner-

Figure 35. Results of the 1st group

Derivative products—mascot on coasters	Commercial packaging—reshaping the image of shell powder

Industry

Figure 36. Results of the 2nd group

Commercial packaging—reshaping the image of drops of essence of clam

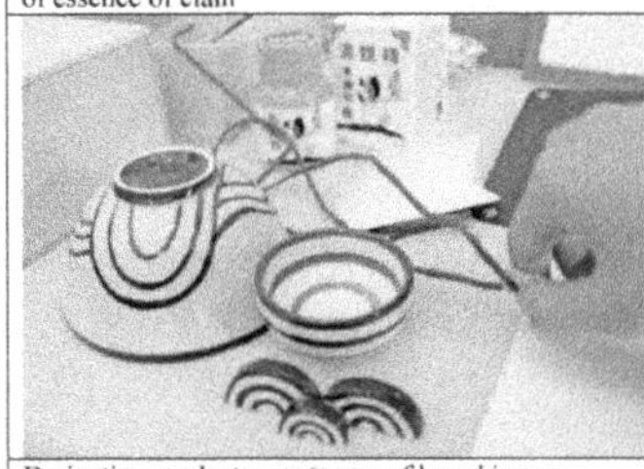

Derivative products—outcome of brand image promotion

Identification of enterprise—CIS

Identification of enterprise—brand image, Tseng

Figure 37. Results of the 3rd group

Identification of enterprise—CIS

Commercial packaging—reshaping of image

Figure 38. Results of the 4th group

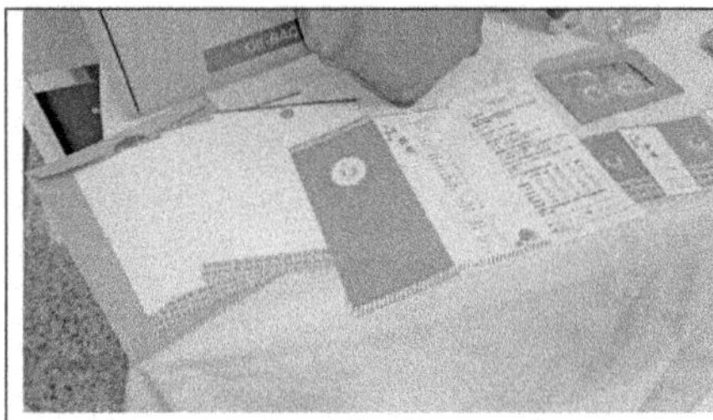	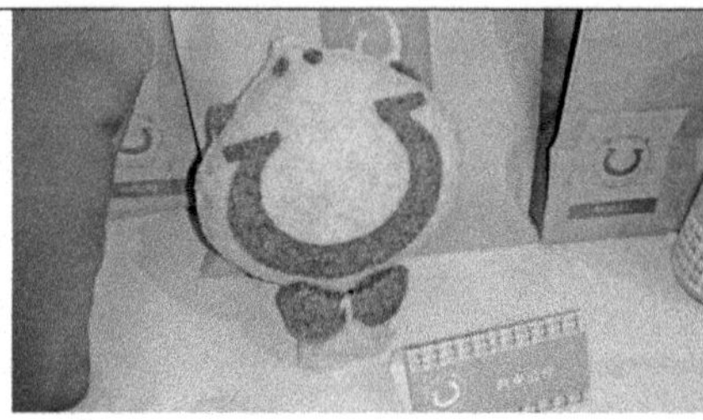
Identification of enterprise—restaurant menu and customer idea form	Identification of enterprise—mascot

ships is worthy of the efforts of universities and teachers; however, in order to gain ideal education results and form education characteristics, it requires personnel and long-term investment.

Industry-academia partnerships can generate benefits and influence in four aspects, i.e., teacher, student, system, and course (Table 2).

Table 2. Benefits and influence of various aspects

Aspect	Benefit	Comparison
Teacher	Increase the frequency of communication with industry insiders to review the gap between curriculum and industry demands, and increase opportunities for further involvement in enterprises	Before: The industry-academia partnerships held by teachers were conducted independently without correlation. After: Teachers have clearer understanding of the close relationship between curriculum and industrial requirements, and are willing to invite industrial professionals to become involved in course design for reciprocity growth.
Student	Increased opportunities for face to face courses taught by enterprise experts	Before: Students had little exposure to insiders of industry for exchange and learning. After: Increased opportunities of exchange between students and industry professionals, application of knowledge in practice, and increased internship and employment opportunities.
System	Introduction of enterprise experts to regularly participate in course teaching	Before: No opportunities for long-term teaching by enterprise experts. After: Enterprise experts have been added as key members with broad participation in future curriculum planning, implementation, and teaching in universities.
Course	Deep conjunction of courses with demands of enterprises	Before: Few chances to participate in discussions with enterprise experts, and a gap between teaching content and actual demands of enterprises. After: Through curriculum reform, the demands of enterprises and curriculum contents are closely connected.

It has been found through the comparison between the differences of the above aspects that the impact on teacher, student, system, and course of industry-academia partnership lies in the bridging of the gap between learning and practice, especially by the technology and practical experiences related to industry shared by industrial professionals in courses or projects.

Enterprises want production and development talents, while universities are cultivating students with various and prospective knowledge and ability because working ability, besides technical level, include work attitude, coordination ability, execution, character and moral, communication skill, and so on. Obviously, design education of universities is designed to train talents meeting the requirements of future industrial development and more than merely teaching of technology. Industry-academia partnership shall focus on students work environment adaptability, establish common tacit agreement and common goals, make students more aware of the actual situation of workplace, and let them have the ability to conduct self-inspection of strengths and weaknesses so as to enhance their competitiveness in the workplace. Nowadays, Taiwanese society and campus have freed and divers atmosphere. Students who participate in design education of industry-academia partnership can learn not only industrial design technology, but also accumulate interdisciplinary experiences by participating in interdisciplinary teamwork, cultivate their teamwork spirit, and broaden their innovative knowledge and interdisciplinary technologies.

CHALLENGES AND DIFFICULTIES FACED BY INDUSTRY-ACADEMIA PARTNERSHIPS IN DESIGN EDUCATION

Industry-academia partnership really helps to improve the R&D level of universities (OECD, 1996, 1999), develop the added value of existing industrial technology, and actively carry out R&D of key technology in order to stimulate industrial upgrading and transformation, activate innovative system, and promote economic development. At present, industry-academia partnership in Taiwan is mainly driven by the Ministry of Science and technology, Department of Industrial technology, Ministry of Economic Affairs, and Ministry of Education. The industry-academia partnership driven by the Ministry of Science and Technology is implemented with the research funding provided by the government, manpower and equipment invested by universities, and supporting funds and personnel provided by enterprises. The results are shared by the Ministry of Science and Technology, research teams, and contributive enterprises. The projects of Department of Industrial Technology, Ministry of Economic Affairs are mainly about technological development. And the projects of Ministry of Education are mainly about the linkage of education system and industrial circle. The average researches done by universities are basic. For applied research and commercial product development, industry-academia partnership can be adopted to convert basic research results into applied research. Therefore, how to effectively convert and disseminate knowledge to realize knowledge diffusion and play an important role in knowledge-based economy becomes an important topic (Bruneel, D'Este, & Salter, 2010; Caldera & Debande, 2010; Crespi, D 'Este, Fontana, & Geuna, 2011; Chen, et al., 2012).

Universities are key organizations to establish research energy for industry (Freeman, 1987; Lundvall, 1992; Motohashi, 2005; Nelson, 1993). Researchers can directly transfer their existing research results to enterprises through licensing, technology transfer, strategic alliances, spin-off firms, and so on. Thus, universities have become the source power for the development of high-tech industry and regional economy (Feldman & Florida, 1994; Powell, 1998; Rosengrant & Lampe, 1992; Saxenian, 1994). From the above, during the industry-academia partnership in design education, enterprises shall know how to use the knowledge exploration and application capabilities of universities to achieve the desired objectives and improve the performance of industry-academia partnership. Most courses related to industry-academia partnership are based on technology and manufacturing model. The importance of practice is rising. Through the support of theories and important literature of academic research, combined with more innovative knowledge, approach, industry-academia verification in practice, such courses will play more important role in specialization of universities (Chen, et al., 2012).

Timpaue (1983) pointed out that between universities and enterprises, there were three urgent difficulties:

1. Enterprises considered universities as educational institutions that had no response to problems, criticism, or wasted resources, and had little understanding of the actual situations of industries. Such misunderstandings could be eliminated by co-planning industry-academia cooperative programs.
2. Enterprises could not understand the benefits of industry-academia partnership and failed to see the long-term reciprocity.
3. Universities engaged in over-defense against enterprises and failed to utilize the support from and cooperation with enterprises.

Huang (1998) considered that the obstacles of industry-academia partnership were, as follows:

1. The goals of technology R&D of universities and enterprises were different, and tended to cause bottlenecks during execution;

2. Professors at universities had neither the time nor the energy to cooperate with enterprises, or they transferred technology and investments to set up their own companies. Incentives were inadequate;
3. Auditing personnel were insufficient, or rewards for professors were mostly related to academic research;
4. There was lack of effective marketing channels in order to accelerate technology diffusion and transfer, or partners were difficult to find;
5. There were tedious and complex requirements for application qualifications and procedures (e.g., at most, a professor could only participate two cooperative programs), which prevented enterprises and professors from going forward;
6. Enterprises that made investments and researchers could not use research results freely;
7. The academic circle lacked cross-organizational and cross-disciplinary talents, or had limited expertise and resources to meet the needs of enterprises;
8. Most enterprises in Taiwan were small and medium, thus, they lacked profound funds for long-term investment.

Universities and enterprises do not have the same operational philosophy or objectives. Enterprises hope to gain profits via products and services, while universities focus on the communication and diffusion of knowledge and creative ideas. Thus, both sides have fundamental differences in terms of cooperation goals, cooperation value, and ownership of intellectual property (Peters & Fusfeld, 1982; Li, 1998). From the standpoint of enterprises regarding cooperation with universities, they found it difficult to understand the benefits of such cooperation if they failed to view it as a long-term vision (Timpaue, 1983; Tseng, 1990).

On the other hand, Taiwan's existing educational system places emphasis on education degree rather than practical experience. In addition, the positioning of teachers at universities remains limited to teaching and research, and there was a lack of incentives for services, such as industry-academia partnerships. Thus, limited by the factors of faculties and funding, as well as a lack of relevant support and rewards, teachers failed to actively participate in industry-academia partnerships. (Lin, 1991; Wang and Chen, 1997) In recent years, the Ministry of Education considered the patents of teachers, technology transfer, and industry-academia partnerships at the same level as papers, hoping to solve the problem of inadequate incentives (Chen, 2000).

Based on the opinions of the above scholars, universities, and enterprises, and due to their different operational and teaching philosophies, there were different understandings of the goals and values of industry-academia partnerships, meaning a lack of communication and mutual trust easily led to obstacles of such partnerships. The effect of industry-academia partnerships shall play a vital role in the recognition of value for both sides (Hsu, 2013).

FUTURE RESEARCH DIRECTIONS

Through cooperation between universities and enterprises, teachers can convey the latest industry information to students as they accumulate practical experience and strengthen the understanding of the status quo and demand of professional knowledge, which not only improves the quality of teaching, but also ensures that students can adapt to the skills required in the workplace. From learning fundamental design to the application of skills, balanced development between practical experience and professional design knowledge and theories can help students understand the truth of the concepts of design, and the real purposes of design education. Moreover, students learn how to cooperate with others during practice related to their design courses, expand their knowledge and skills to learn, and improve their academic performance.

Specifically, universities, enterprises, and students have their respective and specific functions for the future development. Universities should continually emphasize "R&D industrialization" and "practical teaching". On the other hand, enterprises should utilize human resources of universities to promote "R&D improvement" and "transition upgrading" and enhance competitiveness. And, students should shorten the gap between knowledge and practice, expose to market functions, and enhance career competitiveness in the future. In other words, Industry-academia partnership in design education should aim to obtain win-win and mutually reinforcing results sharing. Enterprises should intend more to use the human resources of universities with design and R&D energy to innovate their technology or management processes, enhance operational competitiveness and performance, and obtain suitable talents. Teachers can get R&D resources, improve R&D quality and quantity, expose students to practical experience and technology, and increase students' career ability and employment opportunities.

Obviously, design education of universities is designed to train talents meeting the requirements of future industrial development and more than merely teaching of technology because working ability, besides technical level, include work attitude, coordination ability, execution, character and moral, communication skill, and so on. Industry-academia partnership should focus on students work environment adaptability, establish common tacit agreement and common goals, make students more aware of the actual situation of workplace, and let them have the ability to conduct self-inspection of strengths and weaknesses so as to enhance their competitiveness in the workplace.

CONCLUSION

Industry-academia partnership in design education can not only find research talents required by enterprises, but can also save a great amount of funds for devices by sharing devices among partners. For universities, cooperative R&D can give full play to labor and material resources, and promote economic benefits. Meanwhile, teachers who participate in such cooperation can blend their experience in teaching to optimize the result of teaching. Overall, industry-academia partnerships in design education can help enterprises improve their technologies, and universities improve their research level and train students with the skills required for their future work.

Today, Taiwanese public and private universities and science and technology universities shoulder the responsibility and obligation of diversified development of Taiwan. The participation in industry-academia partnership of students led by teachers aim to link the existing R&D energy of universities with enterprises, help enterprise establish core competencies, commercialize research results, and reduce the risk of early R&D stage. Besides, teachers and enterprises cooperate to develop forward-looking technology or knowledge, enhance future industrial competitiveness, and develop the key technology or mainstream specifications to be applied in leading and new products through high-risk, high innovation, or long-term R&D industry-academia partnership. This confirms that the additive and multiplicative effect of the innovation, knowledge, ability, and application of academic research of teachers can help enterprises transfer useful technology or design and develop products which can be commercialized according to the current market demands on products. In other words, through the journals published or invention and patents received, teachers can convert the new knowledge or new technologies of universities into practical and commercialized output so as to enhance the substance performance of industry-academia partnership. Furthermore, in practice teaching, teachers allow students to absorb innovative knowledge and have practical experiences.

In today's trend of pragmatic, diverse, and intelligent industry-academia partnership, government departments are actively promoting industry-academia programs, establishing education model in consideration of both education and employment, focusing on both theory and practice, and bridging the gap of talents required by key industries. Through the close industry-academia interaction, talents are cultivated jointly. Enterprises can participate in the course planning of universities to carry out practice teaching and provide relevant equipment, internships, or employment opportunities for students. Through the interaction with enterprises, universities can update their teaching materials, provide research equipment for mutual use, and cultivate practical talents via the approach of "learning while doing and doing while learning". The promotion and implementation of "industry-academia partnership", for universities and teachers, is an important direction worth of continuous efforts. However, in order to obtain ideal results or make it a characteristic of the universities, a lot of manpower and time and long-term business are essential. Today, as of university education, students, parents, and universities are supposed to have different ways of thinking. Since there are so many universities, the value of university education shall be redefined. In recent years, many universities strive to transform to get rid of the traditional thinking of "ivory tower" and expand their cooperative partners to public and private institutions, non-profit institutions, or even foreign universities or organizations. Therefore, the attention attached to industry-academia partnership by Taiwan does not lag behind that of Europe or the United States. Taiwan has complete institutions and mechanisms to promote the partnership. However, to further and substantially increase the performance of industry-academia partnership, both system and practice shall be taken into consideration. For instance, in terms of system, we shall think about how to make the multiple leading systems of the government become more effective and forward-looking pushing systems. In terms of practice, the incentive for teachers to implement industry-academia partnership and the attraction of enterprises to invest in industry-academia partnership shall be planned and implemented.

In particular, in the process of globalization, the challenges faced by Taiwanese government and civil society will become more and more harsh. Thus, Taiwan must first recognize the several important traits of globalization and liberalization:

1. The trade liberalization under WTO framework was removed due to trade barriers. Though the stage for enterprises becomes bigger, it also means that enterprises are facing increasing competition. Enterprises must face the competition from both Taiwan and the world.
2. Along with the liberalization of international capital movement, most countries have relieved their capital control. Financial assets volatility becomes more apparent. The effects of crises such as financial turmoil are spread more rapidly with increasing impacts.

3. After the disintegration of communism, more and more Third World countries join capitalism, resulting in the elimination of labor-intensive industries and industries with low entry threshold.
4. Globalized capitalism promotes enterprises to product their products in places with the lowest cost and sells their products to the places with the highest profits.

While searching for production base globally, multinational enterprises are transferring new scientific technologies. To obtain such technology transfer, a pre-requisite is good and sufficient human resources. Based on the understanding of the characteristics of globalization, we can see that the quality of talents plays the most important role in global competition. Taiwan has average advantage in terms of the quality of manpower. However, it no longer has the price advantage of human resources. Thus, no matter Taiwanese government or private sectors shall further enhance the quality of education to establish the basic force of knowledge economy.

During the implementation of industry-academia partnership in design education, enterprises can find suitable R&D talents via the partnership and save huge equipment cost with the support of the equipment and instruments of R&D institutions. For universities, joint R&D can make full of human and material resources so as to enhance economic benefits. Also, teachers who participated in the partnership can introduce their experiences in teaching, combine theories with practice, and obtain more satisfactory results of teaching and learning. All in all, industry-academia partnership in design education can contribute to the technical rooting and new product development of enterprises, enhance universities' R&D level, and cultivate students' career ability for future employment.

REFERENCES

Bloedon, R. V., & Stoke, D. R. (1994). Making University-industry Collaborative Research Succeed. *Research Technology Management*, *37*, 44–49.

Bottge, B., & Osterman, L. (1998). Bring the Workplace to Classroom. *Educational Leadership*, *55*, 76–77.

Brown, T. (2008). Design Thinking. *Harvard Business Review*, 84–94. PMID:18605031

Bruneel, J., D'Este, P., & Salter, A. (2010). Investigating the factors that diminish the barriers to university-industry collaboration. *Research Policy*, *39*(7), 858–868. doi:10.1016/j.respol.2010.03.006

Caldera, A., & Debande, O. (2010). Performance of Spanish universities in technology transfer: An empirical analysis. *Research Policy*, *39*(9), 1160–1173. doi:10.1016/j.respol.2010.05.016

Chang, K. M. (2000). How to Promoting University-industry Cooperation to Enhance Design Capability. *Symposium Proceeding on Design Education of Japan and Asia*. YunTech.

Chen, H. C. (2001). The Practical Method of Industry-academic Cooperation. *Journal of Technological and Vocational Education.*, *61*, 57–61.

Chen, K. S., & Lin, T. Y. (1990). OEM? Or own brands? *Taiwan Economic Research Monthly*, 134.

Chen, T. C., Lin, B. W., & Chu, M. A. (2012). Knowledge Exploration and Exploitation in the University-Industry Collaboration. *NTU Management Review*, *22*(2), 59–86.

Chen, Y. C. (2007). *A Study on Business Strategy of Small and Medium Enterprise in Taiwan for the Transformation of OEM/ODM to OBM*. (MS Thesis). Department of Industrial Management National Taiwan University of Science and Technology.

Chen, Z. H. (2000). The Review and Improvement of Industry-academia Cooperation in Research Program. *Science Development*, 921-927.

Chou, T. D. (2013). Taiwanese Strategic Mode to Promote Industry-academia Partnership—Industry-academia R&D and Talent Cultivation. *The Journal of Chaoyang University of Technology*, *18*, 85–109.

Chung, C. J. (2015). *An Analysis of Key Factors for University Faculty to Participate in Industry-academic Cooperation– Applying Analytical Hierarchy Process Method.* (MS Thesis). Graduate School of Visual Communication Design, Kun Shan University.

Crespi, G., D'Este, P., Fontana, R., & Geuna, A. (2011). The impact of academic patenting on university research and its transfer. *Research Policy*, *40*(1), 55–68. doi:10.1016/j.respol.2010.09.010

Dai, X. X. (2000). *The Popularization and Marketization of Higher Education.* Taiwan.

Feldman, M. P., & Florida, R. (1994). The geographic sources of innovation: Technological in frastructure and product innovation in the United States. *Annals of the Association of American Geographers*, *84*(2), 210–229. doi:10.1111/j.1467-8306.1994.tb01735.x

OECD. (1999). *University research in transition.* Organization for Economic Co-operation and Development.

KEY TERMS AND DEFINITIONS

Cooperative Education (Or Co-Operative Education): This is a structured method of combining classroom-based education with practical work experience. A cooperative education experience, commonly known as a "co-op", provides academic credit for structured job experience.

Customer Relationship Management (CRM): It is an approach to managing a company's interaction with current and future customers. The CRM approach tries to analyze data about customers' history with a company, in order to better improve business relationships with customers, specifically focusing on retaining customers, in order to drive sales growth.

Design Education: This is learning how to apply practical methods, prior knowledge, and natural talent to solve new problems.

Design Thinking: This is people-oriented design spirit and method, taking people's demands, action, and the feasibility of technologies or business into consideration.

Industrial Ecology: This is concerned with the shifting of industrial process from linear (open loop) systems, in which resource and capital investments move through the system to become waste, to a closed loop system where wastes can become inputs for new processes.

Interdisciplinary: It involves the combining of two or more academic disciplines into one activity (e.g., a research project). It is about creating something new by crossing boundaries, and thinking across them.

Original Brand Manufacturer (OBM): This is typically a company that sells an entire product made by a second company or including a component thereof from a second company sources as its own branded product.

Chapter 8
Play, Think, Design:
Play as a Means to Acquire and Enhance Design Thinking Skills

Rémi Leclerc
The Hong Kong Polytechnic University, Hong Kong

ABSTRACT

Designing requires the simultaneous application of diverse modes of thinking and making; the administration of a set of competences, among them creativity, ranked most important by Hong Kong Polytechnic University School of Design faculty members. An introductory design thinking subject offered by the School uses Play as a means to foster best design learning experiences. It leverages Play's integrative cultural agency to facilitate students' acquisition of creative and instrumental thinking skills. From the formulation of a hypothesis and identification of a context, to subsequent iterative development and testing of a design proposition, the acquisition of fundamental creative design thinking skills are facilitated by play and demonstrated over the course of a project. A survey questionnaire (N=219) and subsequent factor analysis revealed positive student feedback. This chapter describes how the subject was implemented, and suggests how this blend of international cultural influences informing design education could serve China's creativity and innovation industries.

DOI: 10.4018/978-1-5225-0911-0.ch008

INTRODUCTION

Designing requires the simultaneous application of diverse modes of thinking and making; the administration of a set of competences (thinking) and skills (making). Among these is creativity, ranked most important by Hong Kong Polytechnic University (PolyU) School of Design (SD) faculty members in the Advertisement, Communication Design, Environment & Interior Design, and Industrial and Product Design Disciplines in a pilot study survey carried in 2012/13. PolyU also promotes creativity as part of certain 'soft skills' it lists in its strategic goals.

Design is often associated with creativity, and creativity in design with divergent thinking. Hennessy and Amabile's conceptual definition of creativity (1988) is that 'a product is considered creative to the extent that it is both a novel and an appropriate, useful, correct, or valuable response to an open-ended task'. Amabile also points that experts can recognize creativity operating in their domain when they see it (in Baer, 1994), a common approach in design education. For Mayfield (2011), innovation is creativity applied to solve a specific problem. It can be viewed as occurring on the individual, team, organizational, industry, and societal levels.

An introductory Design Thinking subject offered by the School leverages Play's integrative development and cultural agency as a means to facilitate students' acquisition of creative and instrumental design thinking skills and foster best design learning experiences. From the formulation of a hypothesis and identification of a context, to subsequent iterative development and testing of a design proposition, the acquisition of fundamental creative design thinking skills are facilitated by play and demonstrated over the course of a project. A survey questionnaire (N=219) and subsequent factor analysis revealed positive student feedback. This chapter describes how the subject was implemented, and suggests how this blend of international cultural influences informing design education could serve China's creativity and innovation industries.

BACKGROUND: PLAY, CULTURAL DEVELOPMENT, AND DESIGN - CONCEPTUAL BRIDGES

Play in Developmental Psychology

Over the last two centuries, the study of play has recognized play to be intrinsic to the human experience; an essential part of humans' and many animals' biological (individual) and social (group, cultural) development and survival. Although offering different perspectives on how social contexts shape the content and process of play, psychologists from Gross in the late 19th century (1896, 1899), to Parten

(1932), Piaget (1959), Vygotsky (1934, 1978), Pellegrini, and Smith (2005, 2009) in the 20th, demonstrated the need for children to engage in play, through stages of physiological, social, and cognitive development, as an initiation and preparation for later life, condition for welfare, and possibly evolution.

Play in Culture

Beyond developmental psychology, Huizinga (1938) argued for an appreciation of play as an agent of civilizational or cultural development, albeit joining his peers in recognizing the difficulty in defining or ascertaining its functional significance. Caillois (1961) redirected an emphasis on competition in play to expand its social significance and offered a model to explain human behavior. Conscious of the difficulty in defining the field, he proposed an open, modular model of play, adaptable to define different ludic experiences. His model includes four fundamental play forms: agon (competition), alea (chance), mimicry (imagination), and illinx (vertigo, or sensory distortion), to employ on their own or in any combination to define play activities. These forms of play could be placed on a continuum opposing ludus (rule-governed, skill-requiring game play), to paidia (unstructured free play). This continuum could also be interpreted as a dialectic characterizing the instability of culture, observing that in time paidia eventually turns into ludus, with rules giving way to the forces of paidia, and vice-versa, in an iterative historical development alternating rule-forming and re-forming. The continuum bears similarity with the divergent (exploratory, creative, expanding, 'right brain')/ convergent (discriminating, analytical, integrative, 'left brain') dialectics observed in the design practice.

Play: Limits and Opportunities for Design

Attempting to address inconsistencies within the field of play, Sutton-Smith (1997) lists seven rhetorics of play, seeing it being adopted as 'a persuasive discourse or implicit narrative... by members of a particular group affiliation or discipline to lend validity to their beliefs and interpretations'. He concedes play is ambiguous and difficult to define. His rhetorics may not explicate play in all its form, and the study of play should encompass children and adult forms of participations. He suggests another form of dialectics: biologically, play activities function to 'reinforce the organism's variability in the face of rigidifications of successful adaptation (...). This variability covers the full range of behavior from the actual to the possible'; and 'psychologically, as a virtual simulation characterized by staged contingencies of variation, with opportunities for control engendered by either mastery or further chaos. Clearly the primary motive of players is the stylized performance of existential themes that mimic or mock the uncertainties and risks of survival and, in so doing, engage the propensities of mind, body, and cells in exciting forms of arousal.' He concludes that despite his extensive criticisms of the rhetoric of progress (i.e. play supporting development), he has invented yet another form of it, reducing it to the 'potentiation of adaptive variability'.

Play and Design

While Sutton-Smith sees play's study evolving into forms of rhetoric, Margolin (2007) suggests that 'as creators of models, prototypes, and propositions, designers occupy a dialectical space between the world that is and the world that could be. Informed by the past and the present, their activity is oriented towards the future.' The creative act of design is a progressive form of discourse: a 'line making a point', a form of purposeful play. Beyond Simon's (1969) consensual definition of design as the 'transformation of existing conditions into preferred ones', design *shapes* culture: Margolin (2009) affirms that 'there is no human activity that is not embedded in material culture'. Everything around us has been designed, from the spoon to the stool, the window to the sill, bikes to buses, haircuts to holidays, and the services - or 'experiences' - we consume to navigate an increasingly artificial reality. Considering this more holistic, accurate, and less utilitarian definition, in the light of Huizinga's contention that culture is the outcome of play, one could easily perceive the links existing between play and design.

Bridging Play and Design

Table 1 highlights conceptual bridges between characteristics of play drawn from Huizinga, Caillois, and Sutton-Smith, with fundamental design.

Play, a Formative Experience for Critical Thinking and Aesthetics

Some may see the early challenges Socrates imposed upon his disciples in the 5th century BC - among them Plato - as a form of rhetorical play. In the sacred Indian Veda Vyasa (8th century), play was seen as an essential component of ontological development. Much later, in his treatise on education 'Emile', Rousseau (1762) suggested play was conducive to acquiring life competencies and applied skills. Inspired by Rousseau, Schiller (1794) suggested that play was the only opportunity for humans to fully develop their humanity by setting free the two aspects of its double nature: sensation and thought… an illustration of the left-brain/side-brain complementarity core to design thinking. Kindergartens creator Froebel devised in 1837 a number of 'Gifts', or basic toys to nurture children's learning through play;

Table 1. Conceptual bridges linking play to design

<table>
<tr><th colspan="2">Play and Design - Conceptual Bridges</th></tr>
<tr><th>Play... ...is...</th><th>...Design</th></tr>
<tr><td>A voluntary engagement of body and senses, arousal</td><td>An engaging practice, embracing the creative act, arousal, challenge</td></tr>
<tr><td colspan="2">Enjoyment of activity, gratification sought</td></tr>
<tr><td>Separate from life routines occupying its own time and space;</td><td>The project, a conjectural problem space, where real-life is reinterpreted</td></tr>
<tr><td colspan="2">Huizinga's 'magic circle', temporary world within ordinary world, time and space needed to focus, and be in creative flow (Csíkszentmihályi, 1996)</td></tr>
<tr><td>Governed by rules that suspend everyday laws players must follow</td><td>Propositions are elicited by a brief, innovation is a self-disciplined process</td></tr>
<tr><td colspan="2">The magic circle may be formally structured, rule-governed, possibly a goal or a solution, determined by social and cultural contexts, and its borders with 'real life' porous.</td></tr>
<tr><td>Make-believe confirm existence of imagined realities that may be set against 'real life', potentiation of adaptive variability'</td><td>Research through design deploying user-centered scenarios, empathetic role play eliciting desirable futures</td></tr>
<tr><td colspan="2">Narrative play, generative, iterative development, 'what if' scenarios elicited</td></tr>
<tr><td>Ambiguity, results not pre-determined, prompting players' involvement</td><td>Design processes' uncertainty, fuzzy front end, 'wicked' problems (Rittel, Weber, 1973)</td></tr>
<tr><td colspan="2">Motivation for discovery and creativity</td></tr>
<tr><td>Unproductive, creating no wealth, aimless, and ends as it begins</td><td>Experimentation, drifting, tinkering, error, rest, dream</td></tr>
<tr><td colspan="2">While design practice is still mostly solution-focused and aimed at production, design schools are privileged environments where experimentation, mistakes, as long as they are done 'in earnest'</td></tr>
</table>

an early from of constructivist education focused on activities in which manipulation of toys was central.

Gropius, Rand, Brown... Designers Making the Case for Play

Designers and artists were quick to pick up on play's constructive relevance to the development of aesthetic and creative thinking competencies. Kandinsky and Gropius in Europe, Frank Lloyd Wright in the US are among a number of designers who recognized the influence of Froebel's Gifts in their development as designers and architects. Visual designer Paul Rand (1965), quoting academic Highet (1950), reminded us that 'the best Renaissance teachers, instead of beating their pupils, spurred them on by a number of appeals to the play-principle. They made games out of the chore of learning difficult subjects', and made a case for a similar 'magic circle' of play in design education, noting that 'a problem with defined limits, implied or stated disciplines which are, in turn, conducive to the instinct of play, will most likely yield an interested student and, very often, a meaningful and novel solution.' Rand's interpretation of play for design should be seen under the light of clearly stated rules (so that students are given ground to play upon), alternate engagement between theory and practice (so that experiential play is informed by principles), and the discipline necessary to design solutions characterized by an economy of means (the condition necessary for good design). Munari, who saw 'Design as Art' (1966), inspired by Piaget's active discovery learning and Montessori's constructivist methods leveraging design and art, created his own 'Giocare con l'Arte' (Playing with Art) educational method in 1977, advocating 'non dire cosa fare ma come' ('to do in order to understand') comprising three consecutive actions: *observe*, *make*, and *reflect on the making*. In 1998 Manu suggested that 'beyond its environmental imperative sustainable design will have to be engaging, challenging, rewarding, absorbing, non-frustrating, and of repeat experience value' - as in play. Design consultancy IDEO president Tim Brown (2008) promoted play's relevance to the design practice in a Ted talk. Kristiansen (2014) leveraged narrative play and manipulation of objects in corporate offices using Lego elements to create tangible semantic props to illustrate prospective scenarios through successive narrative frames.

Play for Design Thinking: Play Enhancing Creativity and Innovation Education

For many, 'life is play'. Yet while it is easy to see play everywhere in human activity, it is not to substantiate its relevance to the creative practice and innovation. Vygotsky's theory of creative imagination (Smolucha and Smolucha, 1986) is constructed of four tenets; firstly that imagination is the internalization of child's play, imagination is directed consciously, creative thinking involves the collaboration of imagination and thinking in concepts, and artistic creativity requires such collaboration between imagination and thinking in concepts.

In a designed world, Debord's Society of Spectacle (1967), where hyperreality trumps reality (Baudrillard, 1981), what you see is not what you get: play and design are omnipresent yet evanescent, and products and images convey multiple layers of meaning, mediating social dynamics and shaping culture. Play and design share many similarities, including the difficulty in being defined. Design like play is a system of communication adults engage into, which is ambiguous - *if you can't take uncertainty, get out of the Studio*. While the uncertainty in design may be unsettling for young students, acceptance of it through the ambiguous nature of play makes it palatable. Hence the case for play in education is intertwined with creative, experiential construction of knowledge – engaging, active, playful forms of learning. While reference is made to play theories and toy design processes as prototypes for design thinking, one's awareness of play's open and ever-evolving nature with respect to its process is - just as in design - paramount to keeping the approach alive and contemporarily relevant. In the 20th century alone, the work of social scientists such as Vygotsky (1934), Huizinga (1938), Piaget (1959), Caillois (1961), Sutton-Smith (1997), Smith and Pellegrini (2005), to name just a few, has shown play, if not an ambiguous concept, is at least one that is defined differently in different sociocultural contexts. Nevertheless, most argue for universality in the need for play and its value. Differences are highlighted in its expression and adoption in different cultural contexts.

Play's Rightness, Design's Appropriateness

As they negotiate an increasingly knowledge-based and prominent profession, engaging in 'designerly ways of knowing' (Cross, 1982), designers integrate into their practice multiple disciplines. Cross argues for Design as an emerging 'third area' of education, an integrative way of acquiring knowledge and generating meaning, filling the cognitive gap left by the shortcomings of science and humanities education. This implies a reorientation from the instrumental aims of conventional forms of knowing, towards intrinsic values such as 'appropriateness', in contrast to 'truth' (in the sciences), or 'justice' (in the humanities). 'Appropriateness' qualifies design solutions more attuned to the immanence of contemporary society. Designers provide solutions that are right for 'now' (designers provide appropriate solutions for present contexts), and 'right' for now (knowing that appropriateness is contextual hence subject to change, solutions for future contexts need improvement or replacement). As design shapes culture, its processes are characterized by uncertainty and immanence. Here parallels could be drawn with play as agency for culture. Sutton-Smith (1997) suggests play's quirkiness (challenging conventions), redundancy (trying out 'what if' scenarios), and flexibility (adapting to circumstances) might nullify the rigidity (a tendency to be satisfied by the *one* solution) that sets in after successful adaption (or designs), thus enhancing human's adaptive variability.

HOW DOES THE DESIGN PROCESS PLAY OUT? PLAY FOR DESIGN: DESIGN PLAY, ENHANCING CREATIVITY EDUCATION

Visualizing Design Play

This paper aims to demonstrate how play enhances design thinking education experience. Designers commonly 'use non-verbal, graphic/spatial modelling media' (Cross, 1990). So how to visualize 'Design Play' processes? In order to strengthen

Figure 1. Visualization of a design process

the case made for play's appropriateness as a platform to explicate design thinking's intricacies, and engage stakeholders into appreciating its nature, a number of models were generated. Figure 1 visualizes the uncertainty and immanence of design processes, from their fuzzy front end, and 'wickedness' of their development, to the resolution of tensions and generation of a solution 'appropriate for now' (until a better one comes along, hence the exclamation 'solution' mark leading up to a new question mark). Dots represent methods and tools deployed.

Conceptual bridges can be highlighted as a 'friendly Trojan hobby horse' to mediate design education and reveal design's relevance to culture. These may be further revealed by visually juxtaposing essential play and design concepts. To such an effect the twofold temporality of design projects will be used as a metaphor: design projects typically follow 'diachronic' narratives, the unraveling of a process. Projects follow a planned, yet evolving course of actions, involving successive 'synchronic' activities, deploying methods or tools. The diachronic dimension is represented in the Design Play Continuum (Figure 2, Leclerc & Wan, 2012); its synchronic counterpart tabled in the Design Play Taxonomy of activities (Table 2, Leclerc & Wan, 2012) revealing the intrinsic playful nature of design.

Design Play Diachronics

The Design Play Continuum reveals the playful narrative nature underlying design processes. Simultaneously working at defining problems and solutions, designers engage in iterative creative processes, alternating between:

Figure 2. Design play continuum

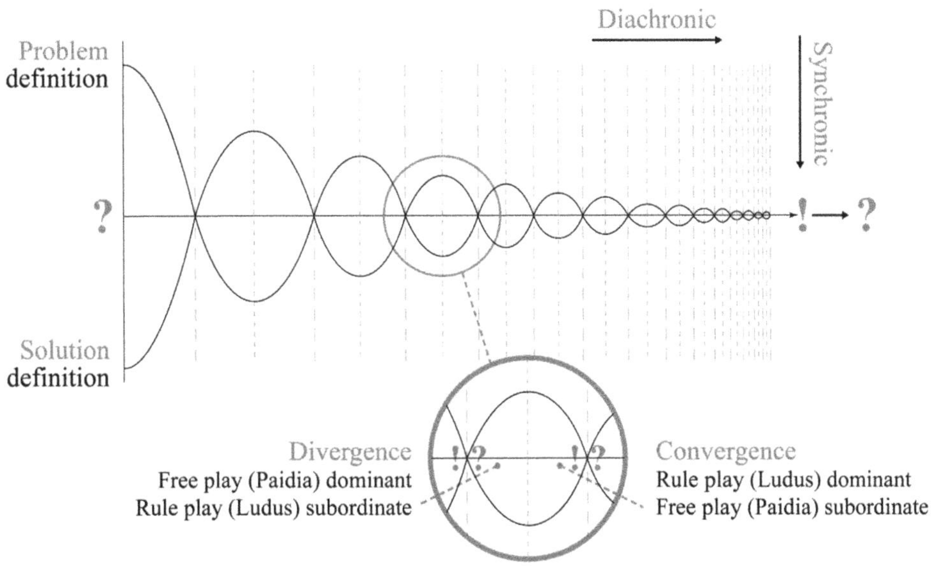

1. Divergent, open phases of discovery, where freer forms of creative play - Paidia - dominate, and
2. Convergent, structured phases, during which analytical and synthetic decision-making akin to rule-based, skill-requiring game play - Ludus - prevail.

Design projects thus play out in a dialectical continuum, alternating between divergent, Paidia-dominant, and convergent, Ludus-dominant phases, ever refining a problem/solution (or question/proposition) until an 'appropriate' formulation of a solution is harmonized in response to a brief, itself the expression of a sociocultural context. Here, creative 'fun' and critical 'seriousness' alternate along a problem-solution co-evolving experiential continuum. The idealized linearity of the process is an ironed-out visualization of its tangled, meandering reality.

Design Play Synchronics

The Design Play Taxonomy lists a summary of play activities mirroring one of design tools and methods so as to reveal design practice's inherent playful nature. Its structure loosely follows developmental patterns, from sensory motor to cognitive

Table 2. Design play taxonomy

Activity	Forms of Activity	Play and Design Activities and Methods	
		Play	Design
Play and design	**Competition** Pushing/ challenging oneself/ others, sensory distortion, testing limits, subversion, empowerment	Contests Games Games of strategy Completing a game Achievement	Competitions Project bids Design strategy Completing a project Promotion
	Cognition Learning, defining problem-solving, rule game, strategy	Games of strategy Quizzes & puzzles Science kits Tinkering experiments Outdoor exploration	Strategic design Problem-solving Reverse-engineering Research through design Field studies
	Creative Exploration, expression, arts, aesthetics	Drawing Tomfooling Arts & crafts Performance art	Visualizing Brainstorming Drawing & model making Presentation & showmanship
	Imagination Self-social, fantasy, pretending, mimicry, narratives	Story telling Role play Make believe Costume play	Scenario building User empathy Prospection & simulation Fashion & identity building
	Manipulation Fine motor, constructing, making	Model making Construction toys Arts & crafts Tools & stationery	Model making, prototyping Design principles Craftsmanship Tools & machine tools
	Sensory (perception) & Motor (body) Feeling & moving, thrill, pleasure	Sports Sensory play Emotional experiences Altering perception Pleasure	Ergonomics, body storming Gestalt, Kansei, synaesthetics Experience design Blindfolding Aesthetics
	Chance Gamble, weighing the odds	Chance games Gambling Magic	Gambits Opportunity & venture Delight

play, through fine motor, imagination, and creative play. Chance, characterizing one's conception, opens the list. Competition, requiring integration of complex skill sets, including cognition and strategy closes the list. Synchronic design activities are inserted in the Design Play Continuum. While the taxonomy takes a table form, cells are not water tight and forms and functions overlap; design's uncertainty mirroring play's ambiguity.

PLAY BRIDGING CULTURES

Hong Kong's Context

As explained above, leveraging play's developmental (individual, biological) and cultural (group, social) agency helps revealing how designers shape culture. Also, play's engaging and non-threatening nature makes it an appropriate platform to foster constructive bridges between cultures. Unlike the rest of China, HK's tertiary design education environment is multicultural yet ethnically homogenous: educational content is international, context is local. Over 90% of SD's undergraduate students share a common secondary education background shaped until recently by Chinese and British colonial educational traditions, which place emphasis on performance in terminal examinations. About 40% of their teachers come from overseas, while the majority of the 60% Chinese have studied or practiced overseas. Also, contemporary design is set in an international context, its practice shaped by methods devised mostly in Europe or the US. Leading academic design journals informing practice and education are also mainly based there and when not, most editorial board members are non-Asian.

CHC Students: Shifting Cultural Realities

'Cultures are anything but homogenous, all-encompassing entities' (Atkinson, 1999). 'Most scholars in the humanities and social sciences no longer speak of culture as a unified, organic object with independent, causal power'. Rather, culture is made of 'webs of significance' human beings 'spin around themselves' (Zimmerman 2006): meaning and language created and shared by a community. Any attempt at defining the 'Chinese Learner' should be guided by the knowledge that China is undergoing sociocultural changes at a pace never experienced in its history. For Ryan (2010), 'the rapidly changing nature of the Chinese education landscape on curriculum reform as well as pedagogy makes any stereotyped interpretations over these learners inappropriate and out of date'.

Watkins and Biggs (1996, 2001) were among the first to dispel myths about 'Confucian Heritage Culture' (CHC) and the 'Chinese Learner'. For them, 'widely held Western stereotypes and misconceptions of Chinese learners are shown to be largely without foundation'. They describe two aspects to a paradox in the Chinese educational system:

- **Chinese Learner:** Despite perceptions of substandard classroom conditions which, according to Western standards, cannot be conducive to good learning, CHC students' often outperform their 'western' counterparts.
- **Chinese Teacher:** Despite classroom conditions, CHC teachers manage to engage their students in productive learning.

They explain that CHC students' learning is meaning-oriented, and that memorizing and understanding are not separate concepts but are entwined, each being a condition for the other.

The 'paradoxes' stems from two common misperceptions:

1. Confusing rote learning for repetitive learning, where the latter is a deep-learning strategy mistaken as a form of surface-learning; and
2. Misconception about the relationship between teaching methods and learning process.

A generalization of the effect of class size on students' learning in isolation from other systemic factors is misleading. A review of CHC students' learning motivation, attributions for success and failure, and other factors, including differences with international approaches, led Biggs (1996) to define his 3P (Presage-Process-Product) model to create 'good learning environment'.

Salili (1996) reviews literature and determines the following characteristics of achievement among Chinese students:

- Individual achievement based on collective values related to that of family and social life.
- Effort is perceived as the key to academic success over ability.
- Lack of praise and high parental expectations develops strong internal motivation for achievement – potentially stifling socioemotional development.
- An authoritarian, conformist education system may not be conducive to developing creative and analytical thinking.

While social - and hence contextual - factors, the identification of one's own achievements with that of the group, and the focus on 'effort' – a form of contribution to group development signaling devotion to the group rather than personal achievement, is beneficial to design project dynamics, hence to design education, the authoritarian and conformist education system does not prepare its students well for creativity and critical thinking, two competences fundamental to design and innovation. While emphasizing deep learning through repetitive approaches, she argues CHC students learn well through concrete examples, doing better in concrete subjects but not in abstract thinking, which might not be conducive to the development of creative and analytical thinking.

Tang and Biggs (1996) identify several 'backwash' phenomena in HK students' learning assessment experiences caused by the examination system:

- Weakness of qualitative assessment practice in schools;
- Cue seeking culture among HK students;
- Understanding for the course requirement; and
- Target-specific strategies developed by students when coping with quantitative assessment etc.

This pushes HK students to play educational 'games' for the sake of grades at the expense of deep learning, essential to allow adaptation strategies needed when one is faced with novel situations and uncertainty as in design.

So how to nurture creativity and innovation in a multicultural context with Chinese characteristics like that of HK's? And how could curricula thus devised be of relevance to China's design education contexts?

Meta Culturalism, Deep Learning

With novel cultural dynamics shaping contemporary Chinese societies, learning styles should not be derived from elusive cultural chimera but discernable contextual bases, making the emphasis on one's background less relevant than the creation of nurturing learning environments and relationships. Design educators, leveraging the socially inclusive nature of play, could 'step back' and take a 'meta' approach to all cultures (Louie 2005), and consider home and host cultures, their potential benefits and shortcoming, so as to understand them as systems. This broader, holistic view of cultures allow for transcultural appraisal and inclusiveness. For Louie, 'Many students come from rapidly changing cultures and those who succeed in life learn to manage the cultural changes within their own cultures skillfully.'

Biggs' (1989) Presage-Process-Product (3P) model, a tertiary student-centered pedagogical approach privileging deep learning, could provide a framework facilitating mutual engagement between educational actors, social interaction, experientiality, and accommodation of meta-cultural sensitivity as the teacher 'steps back'. A deep approach is characterized by an intention to understand course material using strategies such as reading, combining multiple resources, discussion, and reflection, relating parts to a whole and applying knowledge into real world contexts, a common blend of tactics in design education. The 'Presage' phase involves the interaction of the student and teaching contexts and brings students' learning predispositions into the system, such as prior knowledge, abilities, values and expectations, and ways of learning, allowing them to shape learning processes. Meanwhile teachers set institutional scaffold, through course structure, curriculum content, and teaching and assessment methods. Students interpret the teaching context and construct study strategies they think will lead them to achieving expected learning outcomes devised through dialogue. The 'Process' phase sees students process study, a function of both learner characteristics and teaching factors. The 'Product' phase sanctions the attainment of high quality learning outcomes, such as understanding, independent learning, critical and creative thinking, problem solving and other lifelong learning attributes while a surface approach will result in lower quality outcomes.

Universality of Play

While the play instinct is universally shared by all humans, its expression is shaped by cultural contexts. Differences, once shaped by geo-climatic specificities, are now split along industrialized and developing societies. There are few studies on adult play in transcultural contexts, however Pope Edwards (2000) reports that 'children in all communities (,,,) seemed to have an appetite for self-expression, peer collaboration, exploration, rehearsal, imagination, and problem solving' (all elements core to design practice). Also play and work were not clearly separated, with each facilitating the construction of knowledge, skills and schemes and extent of control over their environments. However she points out cultural norms determined the degree to which play was stimulated by physical and social environments, whether play was a good use of time, or preferred to preserve tradition or instigate innovation, and whether environments provided easy access to models and materials for creative and constructive play. Multiple play types are praticed in different communities and depend on the environment only for reinforcement, not for instigation.

Salen and Zimmerman (2003) lay out the 'interactive, representational, social and cultural aspects (of play) as simultaneously contributing to the experience of play.' Here, 'games are complex forms of designed culture to be understood from multiple perspectives'. Through cooperative, or social, and narrative play, a universal and non-threatening platform could be formed, a transcultural go-between through which people come together and learn. Play creates moments when differences dissolve, fear melts away, revealing what connects us rather than what divides us, facilitating meta approaches to culture.

Chinese learners are highly adaptive (Volet and Renshaw, 1996). As they jump from the structured, teacher-centered secondary school system into that of an international, uncertain, student-centered educational context (a design school), early study is a challenge. 'Chinese students favor one-to-one interaction, or more informal group settings in tutorials' Biggs (1996). With tutorials being a staple of design education and opportunity for intercultural social play, challenges are eased and students quickly adapt to new cultural environments.

HK'S RELEVANCE TO CHINA

Naming, Framing HK

The author, a HK-born European having lived in HK for close to half his life, with a family history in East Asia spanning three generations, is aware of possible bias as an 'external insider'; however in HK's multangular hall of identity mirrors, it is the privilege of all its residents to claim a local sense of identity, given the fast-paced nature of its constant flux (Leclerc, 2005).

Is HK relevant to China as a 'creative hub'? As the favored gateway into China for most of the 20th century, the now Special Administrative Region (SAR), once defined itself and for the rest of the world as a hub of trade and cultural exchange. The territory's singular geopolitical context inspires a host of interpretations as to its role in contemporary China's development, not without controversy, as it precariously stands on the interface of two worldviews. In the years before and after 1997, the prospect of the Territory's retrocession to the central Chinese authority jolted Hongkongers from all circles into defining a sense of identity for themselves – especially with the handover process being given a 50-year transition period with 2047 as a deadline. Its short, exclusively modern history, coupled with a foreboding sense that the 'party' would soon be ending, fueled the imagination of everyone with a stake in the tiny enclave – for many years the world's most densely populated territory. HK has been seen as 'Asia's World City' (HK Government) a 'lab for the future' (Manzini, 2003), characterized by a 'situated unsituatedness, and authentic inauthenticity' (Fry, 2003), a 'method' (Chan Koon Chung, 2004), as a third of the global 'Nylonkong' economic pole (Elliot, 2008), an 'international focal-point' (Steiner, in Clark, 2009), within its own 'Creative Ecologies' (HKdC, 2009).

Moving Forward: Opportunities for HK, a Leading International Design Education and Business Hub

There is in HK an ominous Hegelian feeling that its sense of identity is disappearing just as it self-actualizes; that as the owl of Minerva takes its flight, the dusk has begun to fall. For Clark (2009) HK represents a unique case study:

With the looming uncertainly about its future, Hong Kong needed a hybrid or 'hyphenated' subjectivity that could negotiate the future complexities of globalization, while acknowledging its mixed legacies of colonialism, nationalism, and capitalism. (...) Its state of constant becoming - in fact helped to define the region's unique design opportunity.

Despite the uncertainties over its future, the HK SAR welcomes more foreign residents. The number of ethnic minorities increased significantly by over 30% from 2001 to 2011 (HKSAR Govt., 2011). From a population of contractual expatriates seeing the territory as part of career moves, the international professionals landing in the SAR nowadays are more entrepreneurial, tend to establish families, and see their stay in the long run. They see the value in creating a home base in HK, enjoying quality of life while eyeing on the potential developmental infrastructure the Pearl River Delta (PRD) network offers to entrepreneurs. Also, and significantly for HK's value as higher education hub, the territory was ranked 5th in the world as a place for studying (QS, 2015), the only Asian territory featured in the top five. Also for over 10 years HK has established the largest annual design event in Asia the Business of Design Week (BODW), to which internationally renowned practitioners and academics congregate to share knowledge and inform the local design community, including students, of fundamental trends in design.

HK should also be perceived through the lens of its integration into the broader realm of the PRD region and China, in which it serves as a center of design creativity and professional services, leveraging the PRD's technological capacity to develop and manufacture products for export. Most headquarters of companies associated with international business are based in HK. HK also co-organizes events with counterparts across the border, such as the world's only Bi-City Biennale of Urbanism\Architecture since 2005. Universities establish outposts in Shenzhen, taking advantage of its technological facilities. The Hong Kong Design Centre staged at Shanghai's 2010 Expo the multi-faceted 'Hong Kong: Creative Ecologies - Business, Living, Creativity' to promote HK designers' energy, style, innovation and branding power to China's mainland and global audiences'.

The international relevance of HK to global design education and practice is not new. As early as the 1970s and 1980s Wucius Wong's formalist design handbooks (1973, 1977) were internationally influential. Today HK leads Asia in design education. It boasts the only Asian school ranked among the world's top 25 art and design schools (PolyU SD) according to Quacquarelli Symonds (QS) World University Rankings by subject (Art & Design) 2015. According to that school's website it also places first in academic productivity in Asia. Situated at the confluence of international influences, HK is best placed to generate valuable homegrown creative design approaches for China and the world.

THE INTRODUCTION TO DESIGN THINKING AND PROCESSES SUBJECT CURRICULUM

Introduction

An introductory design thinking course offered by SD to novice design students uses play theory and toy design as paradigms to explain design processes' workings, nurture creative design thinking practices, and foster best design learning experiences. Through creative iteration of a proposition and its subsequent development and testing, the fundamental design skills of analysis and synthesis are demonstrated over the course of a project development. Leveraging play, students explore creativity and organize instrumental thinking, from the formulation of a hypothesis, or the identification of a situation, to the development of a design solution.

Based on a review of design methodology literature, Cross (1990) names a set of eight basic 'core features' of design ability, or skills. Designers...

- Produce novel, unexpected solutions;
- Tolerate uncertainty, working with incomplete information;
- Apply imagination and constructive forethought to practical problems;
- Use drawings and other modelling media as means of problem solving;
- Resolve ill-defined problems;
- Adopt solution-focused strategies;
- Employ abductive/productive/appositional thinking; and
- Use non-verbal, graphic/spatial modelling media.

Dorst and Reymen's (2004) model of the development of design expertise based on Dreyfus and Dreyfus' general skill acquisition model (1980, 2003) lists seven levels of expertise, from 'novice' to master through 'advanced beginner', 'competent', 'proficient', 'expert', 'master'; 'corresponding' with seven ways of perceiving, interpreting, structuring and solving problems. They discuss the general model's benefits and limitations, and relevance for application to the field of design. Emphasis on instruction, as well as level-to-level transitions moving from instructional to experiential skill acquisition modes are reviewed in the process, opening up possibilities for further adaptation of the model in design education.

The design thinking subject is designed for freshmen, i.e. novice students who have just left behind them the conservative, repetition learning approaches employed in HK's secondary school educational system. Consequently most see the design course as a liberating experience, believing they are freed from the competitive, normative educational model they have endured for 6 years. Little do they realize design is not a discipline where 'dreams come true' easily, and how demanding the design practice is. Play's 'student-friendly' nature eases apprehension of such complexity. While it aims at nurturing 'T'-shaped talent, involving deep knowledge in one field and broad contextual awareness and interest in others, the subject does not differentiate 'professional' design thinking skills from 'transferable' ones: design abilities are by nature professional *and* transferable. Echoing Schiller's views on play as the best opportunity to reconcile sensation and thought, Peirce argued that the best way to create new knowledge is to employ abductive or appositional thinking, simultaneously using both our brain's hemispheres. Teaching fundamental design thinking skills to novice apprentices means facilitating acquisition of forms of thinking not deployed in secondary schools. Because play is non-threatening and integrative, it efficiently introduces such principles, while harmonizing instructional and experiential pedagogy.

OBJECTIVES

The Freshman Seminar: Introduction to Design Thinking & Processes subject, a 14-week, 3 credit subject taken by an average 250 students every year since 2012, aims at introducing students to the multiple facets of design, its definitions, and how it contributes to individuals, society and the economy. Tutors guide students through their design study and help them identify personal interests within their chosen design discipline, and articulate a study path for their 4-year undergraduate design study. The seminar emphasizes the critical relevance of contextual research and theoretical argument to conceptual design and iterative development processes. Table 3 highlights the subject's learning outcomes with design thinking functions and play categories.

PROJECT BRIEF

Students are introduced to basic notions of play and their significance to the design practice. An open-ended project brief requires them to design and build out of upcycled materials a plaything (not a toy) inspired by the five senses. Designing a plaything opens up the possibility to toy with a critical appreciation of design and conventions of interactivity. As novice designers grapple with the complexities of design thinking, they are also invited to think about design's purpose and function. Upcycling waste materials confronts them with consumerism's realities, while

Table 3. Learning outcomes' relevance to design thinking functions and correlation with play types

Upon Successful Completion of this Subject Students are able to...	...Demonstrating Basic Command of Base Factors for Thinking Functions as...	...While Engaging in Such Play Categories as...
Define design and substantiate its contribution to society	Design awareness	Cognitive, narrative, strategic play
Distinguish different design discipline practices to identify own areas of interest and articulate personal development paths	Design discipline Design mindfulness	Narrative, imaginative, attunement, social play
Critically analyze ill-defined briefs and employ multiple modes of thinking to identify issues relevant to contemporary and future social contexts	Design clairvoyance	Cognitive, imaginative play
Articulate appropriate project research and development plans, utilize design tools	Design planning Design research	Cognitive, exploratory, strategic play
Manage uncertain, multiple design thinking processes and iterative development	Design strategy Design management	Cognitive, creative, narrative play
Contextualize design thinking within sociocultural realms	Design contextualization	Cognitive, imagination play
Harmonize personal perspectives with those of work team members, empathize with stakeholders	Design empathy	Social, narrative, cooperative, role play
Integrate concepts and design innovative and feasible solutions appropriate for stakeholders	Design innovation Design pragmatics	Creative, construction, social play
Organize effective verbal, written, visual (2D, 3D, 4D) design project communication (including working prototypes and communication systems)	Design communication	Narrative, poetic, showmanship, manipulative, creative play
Rationalize appropriate aesthetic judgment	Design aesthetics	Language, aesthetics, narrative play
Develop personal work methodologies, including critical, creative, reflective thinking, assert self confidence	Design ontology	Subversion, social, narrative, recapitulative play

suggesting that creativity is often serendipitous and opportune; that ideas, forms, features, or innovation, are 'found': the trick is to cultivate an eye for opportunity. The sensory theme provides students with opportunity to appreciate experiential and phenomenological aspects of design: physically or metaphorically playing with senses introduces them to design affordances, experience, and emotion. Designing a plaything emulates the richness of these experiences: it requires ingenuity and an understanding of sensory, cognitive, material, interactive, emotional, and spatial concepts. Meanwhile, play harmonizes inductive, deductive, abductive and appositional forms of thinking.

Project Rundown

Students work in groups of 5-6, and role-play different research and development positions. Experimenting with mindmapping techniques, they appropriate the brief; establish project scope, and play with various tools to research play theory, materials, sensory experience, and users, defining 'plaything' and claiming ownership of the project. IDEO's Method Cards (2003) are introduced to stimulate user research. The open-ended brief compels them to define the project and the notion of a plaything on their own terms: opportunity for design play emerges from ill-defined briefs. Research informs articulation of a project assessment tool, and group theme development, which individual members interpret to develop playthings. Brainstorming's ice-breaking social value and conceptual 'ignitor' is emphasized over its efficiency in producing actual solutions. Creative impasses are negotiated using creative cards (Von Oech, 1992, Eno and Schmidt, 1975) or critical communication (Fletcher, 2001, McLuhan and Fiore, 1967). Critical design (Dunne, 1999), ludic design (Gaver, 2004) and design as art (Munari, 1966) open up perspectives on design interaction. Progress is documented in a 'lablog', a visual diary recording a journey of discovery, successes, failures, frustrations, despair and joy, and imaginative leaps of faith into the unknown (Fung, Lo, Rao, 2005). Students construct working models and multimedia project communication in view of a group collection presentation, where showmanship, visual delight, and narrative play are encouraged to nurture a celebratory closure of the project. Deliverables include models, catalogs, instruction leaflet-cum-posters, the lablog, and a project report. At each stage of the development process students are reminded of the parallels between design and play so as to facilitate integration of design's complexities.

Student Validation

A Principal Component Analysis (PCA) using SPSS was run on a 20-question 'Design Play' survey taken by the subject's students ($N = 219$) in December 2014 to assess whether they found play to be a useful medium to understand, acquire, and practice design thinking and processes (understanding, learning, and practice are part of design education, as students 'learn by doing' in a Studio-based context, and that assessment is continuous. The suitability of PCA was assessed prior to analysis. A factor analysis was performed and inspection of the correlation matrix showed that all variables had at least one correlation coefficient greater than 0.3. The overall Kaiser-Meyer-Olkin (KMO) measure was 0.9 with individual KMO measures all greater than 0.8, classifications of 'Meritorious' to 'Marvelous'. Bartlett's Test of Sphericity was statistically significant ($p < .0005$), indicating that the data was likely factorizable. A rotation in 37 iterations by Varimax revealed six components that had eigenvalues greater than one and which explained 41.3%, 6.2%, 5.5%, 5.1%, 4.8% and 4.6% of the total variance, respectively. The six-component solution explained 67.5% of the total variance.

Table 4. Design Play survey questionnaire rotated component matrix

Rotated Component Matrix							
Questions		**Factor**					
		1	**2**	**3**	**4**	**5**	**6**
Play helps me become **aware** of project **contexts, understand** design **requirements**, and in this way, better **serve stakeholders**	**Q02**	**.710**	.127	.239	.006	.205	-.047
Play helps me **identify** design **problems, conduct** research, and **experiment**	**Q01**	**.633**	.070	.225	.291	.116	.313
Play supports clear, effective **exchange** of thoughts, feelings, and information	**Q08**	**.577**	.188	.145	.236	.027	.451
Play helps me better **plan** and **organize** my designs, and manage my time and my resources	**Q04**	**.549**	.362	-.030	.338	.164	-.072
Play promotes **critical thinking, perception**, and **defining value** in design	**Q18**	.002	**.760**	.097	.197	.152	.289
Play helps me prepare design presentations, and **communicate** efficiently with others	**Q19**	.303	**.640**	.109	.108	.159	.306
Play sharpens my **aesthetic sense**, perceptual capacity, and critical thinking ability	**Q15**	.240	**.581**	.368	.119	.265	-.135
Play helps me **adapt** to changing design requirements, and **restructure** my designs when needed	**Q11**	.309	**.448**	.422	.163	.096	.260
Play helps me to **generate ideas, think outside the box**, be **resourceful**, and **innovate**	**Q07**	.224	-.002	**.733**	.348	.047	.186
Play promotes designers' awareness of contexts, **open-mindedness, sensitivity**, and **feeling for others**	**Q05**	.045	.244	**.657**	.256	.281	.006
Play allows designers to **adapt** to changing design project circumstances	**Q06**	.269	.416	**.642**	.057	-.057	.214
Design is like playing an **open-ended game**	**Q03**	.109	.009	**.529**	-.066	.438	.329
Through play I am able to generate **new, valuable ideas** and designs	**Q13**	.120	.128	.261	**.768**	.168	.036
Play helps me **map** complex information, **identify** topics, and **organize** knowledge for design	**Q12**	.232	.409	.133	**.644**	.050	.184
Play helps me generate **meaningful** design concepts and/or **personalize** my designs	**Q09**	.391	.049	.163	**.555**	.365	.253
Play makes design education **enjoyable**	**Q14**	.479	.138	.164	.069	**.665**	.103
Play is an **appropriate medium** to teach design	**Q16**	-.042	.211	.124	.349	**.662**	.328
Play makes design thinking **easier to learn**	**Q20**	.399	.324	.100	.221	**.634**	.058
Play enables designers to get out of their 'comfort zones', accept **uncertainty, take risk**, and **achieve better results**	**Q17**	.000	.304	.186	.064	.197	**.781**
Play **influences** people to **achieve** better **design outcomes**	**Q10**	.398	.141	.216	.279	.220	**.552**

Initial Interpretation (Based on PCA Factor Analysis with 4 Factors)

Table 5 lists 6 factors revealing how students link design with play. Questionnaire responses indicate students appreciate the value of play in mediating appreciation of the intricacies of design thinking.

Factor 1 shows students realize play helps students explore and congeal design on the onset of a project, helping them become aware of project contexts, understand project requirements, and in this way better serve stakeholders. Such awareness helps them identify design problems, strategize a course of action, conduct research, and experiment. Planning supports clear, effective exchange of thoughts, feelings, and information, to organize designs, thus facilitating better management of time and resources. Factor 2 demonstrates students see play as a valuable framework for building up on planning and exploratory design work, to discriminate and define value, deploying critical thinking, facilitating adapting and restructuring to design requirements. They concede this helps articulate and sharpen their personal aes-

Table 5. Design Play questionnaire factor analysis

Design Play Questionnaire Factor Analysis			
	Value of Play for Design Questionnaire Key Words	**Base Factors for Design...**	**...Supporting Design Functions**
1	Context awareness, understand requirements, serve stakeholders, identify problems, research, experiment, exchange, plan, organize	Explore/ congeal design	Design awareness Design strategy Design planning Design research Design management
2	Critical thinking, perception, defining value communication, aesthetic sense, adapt to requirements, restructure	Discriminate design, perspicacity, rationalize value choices	Design clairvoyance Design aesthetics
3	Generate ideas, think outside the box, resourcefulness, innovation, open-mindedness, sensitivity, feeling for others, adapt to circumstances, open-ended game	Open awareness, sensitivity, game-like approach to changing contexts	Design empathy Design innovation Design contextualization Design pragmatics Design management
4	Generate new meaningful, valuable ideas/ concepts, map information, identify topics, organize knowledge, personalize designs	Make/ organize meaning, create personal/ other value, appropriate project, integration	Design communication Design innovation Design ontology
5	Appropriate enjoyable medium, learning	Improve pedagogy (teaching & learning)	Design ontology Design discipline
6	Accept uncertainty, take risk, influence, achievement, better outcomes	Achieve results (in uncertain conditions, leadership	Design communication Design management

thetic sense, and effectively communicate choice rationales. Factor 3 establishes that students agree play is instrumental in exploiting the critical ground thus set (see factor 2), seeing it necessary to think outside the box and generate innovative ideas. They concur play promotes a resourceful open-minded sensitivity and empathy, allowing them to adopt an open-ended game so as to adapt to changing design project circumstances. In factor 4 students accept through play they are able to manage creativity and map information, identify topics, and organize knowledge. Play enables them to generate new ideas/ concepts, appropriate the project, and harmonize creativity so as to create meaningful, valuable designs, and personalized designs. Factor 5 unveils students' appreciation of play in regard to educational motivations. Students see the acquisition of design skills is facilitated by and made enjoyable through the medium of play. Finally in factor 6 we see students welcoming play as an agent stimulating risk-taking, getting out of one's comfort zones, and accept uncertainty, empowering them to lead teams in overcoming obstacles and influence project 'game-players' to achieve better outcomes (leaders can handle ambiguity and by nature influence others).

CONCLUSION: RELEVANCE OF EXPERIMENT TO CHINA AND EMERGING COUNTRIES

Play and Design

As demonstrated above, play and design share common fundamental features. This helps place design in a broader conceptual context, explicating how it shapes culture, thus contributing to the study of the nature of design processes. This also facilitates the education of design thinking and creativity. Using the 'friendly Trojan hobby horse' of play rationalizes design's complexities by setting a familiar philosophical ground - we connect naturally to play. Play helps assert the purpose of the design practice in a humanistic light, so in their roles as mediators and moderators designers facilitate a meaningful adoption of technology, for…

- Designers temper the affect of technology, and
- Designers tamper the effect of technology.

Designers may be seen as 'play masters', harmonizing sensation (tempering affect) and thought (tampering effect) as they shape culture. Their design practice is a willful participation in an enjoyable and gratifying activity; where the design project is akin to a 'magic circle', a temporary world within the ordinary world, yet beyond routine; where time and space needed is set aside to focus, and engage in creative flow. This conjectural problem space, its borders with 'real life' porous, where the latter is reinterpreted, may be formally structured, rule-governed, and purposefully orientated towards a goal (a solution), determined by social and cultural contexts. While the solution-focused design practice is aimed at production, or 'scaling-up', design schools are privileged environments where experimentation, and exploratory, critical, or speculative design, and mistakes - if done 'in earnest' and resonate with designers themselves - should be encouraged. This sets the ground for the creation of narratives, in iterative user-empathetic 'what if' scenario development, eliciting desirable futures. A philosophical appreciation of design in the light of play - thinking of design as a positive agent of sociocultural transformation - helps explicate design diachronic (continuum) and synchronic (activities) paradigms to design novices.

Research and validation of the utility of play in adult activities are in their early stages. However as Rand, Manu, Brown, Gaver, Kristiansen, and this author have experienced, both education and practice studio experiences are enhanced by the mindful introduction of play. Experts from diverse fields agree that play is a biological and sociocultural transformative agent, and that outcomes of play, if not prescribed, are nevertheless positive. Play's ontological nature (self-actualization of individuals), its relevance as an agent of cultural exchange and development, its harmonization of competencies fostering creativity, enhance the explication of design's processes, and enrich the design activity, asserting its central place in the human experience.

Play Suitable for Navigating the Immanence of Cultures

Design Thinking is a semantic process, a 'conversation' between stakeholders, generating meaning and shaping culture. With the conversation gone online, it is shaped by anyone, anywhere, all the time. The comforting certitudes of cultural edifices shaped over centuries by geoclimatic contexts – the heritage of tradition asserting identity – are now eroded by the winds of a globalised economy and the postmodern cultural condition. The challenge lies in explicating design's ability to navigate uncertainty, as it shapes the hyperreality of a Spectacle that has gone cyber. Play's ambivalent nature, as it 'pretends to pretend' and tests out biological and psychological possibilities, makes it a socially appropriate transcultural framework to teach design's intricacies, and reveal its purpose to novices and enhance experts' practice.

Play's social inclusiveness also flattens organizational hierarchies, a key to creativity, innovation, and developmental success. The benefits of flat organizational structures were empirically demonstrated as early as 1972 (Ghiselli and Siege), suggesting organizations are more productive when workers are allowed to participate in decision and innovation processes.

This paper asserts play's socially inclusive and creative flexibility bridges cultures. Its context is that of HK, where the pace of change and cultural integration is one of China's most intense, more specifically as it enjoyed a relative political stability while being at the forefront of contemporary global economic changes. Based on this research project's questionnaire factor analysis some questions could be improved, but the survey as it has been fielded yields a good indication of how students see the value of play. Results suggest the ontological, socially inclusive, creative, integrative cultural agency, and non-threatening framework of play, makes it easier for Chinese design novices set in a transnational cultural context to approach the complex, challenging, uncertain, and multiple dimensions of design processes. It provides them with an inclusive, student-centered, and nurturing platform from which to foresee, appreciate, and harmonize design's rich cultural implications. To design playthings - a metaphor for the whole of design practice - is to seriously play at making things for others to toy with. For Clarétie the toy industry encapsulates the best of all design practices. As designers toy with affordances, apprentice designers could tear a few pages from the toy design book. The value for design Gropius, Munari, or Rand saw in the play instinct finds resonance today in the growing number of voices, such as that of Manu, Brown, and Kristiansen, supporting the integration of play with design as a means to nurture best innovation practices.

Design Education for Future China

China's education policy has traditionally been shaped to meet national development needs. China is emerging from an export-driven manufacturing economy to one that is innovation-driven, where creativity needs to be leveraged to provide meaning to its future socioeconomic development. Notwithstanding strategic support for hiring established heads of leading US or EU innovation companies to steer the development of innovation in Chinese companies government knows well it is in educating its future innovators at home that true homegrown development will happen. In its 2006 'Medium- to Long-Term Plan for the Development of Science and Technology', the Chinese government declared its intention to transform China into 'an innovative society' by 2020 and a world leader in science and technology by 2050. (…) The most important revolution in Chinese higher education today may not be its size and scope but that even under the leadership of engineers, top institutions have come to understand that an education in the absence of the humanities is incomplete. (…) This is an education revolution within a revolution, the outcome of which is not yet clear.' (Abrami, Kirby, McFarlan, 2014).

Hong Kong is a part of China, developed by Chinese. Albeit late in its rule, the British administration delegated much of its governance to HK Chinese. Benefiting from an overture to the rest of the world, HK's community harmonized international influences to create a unique cultural expression of Deng Xiao Ping's vision of One Country Two Systems. Asserting HK's attachment to the Mainland while demanding it justifies its relevance to the Motherland would be guilty of double standard. The experiences possible only in HK in this context are trailblazing for China, as the SAR's development is set within the central government's development policies since 1997. HK is necessarily a Chinese design Lab for China. CHC, or Chinese learners in HK or China are not significantly different, and any differences are evened out very quickly in the studio. Given the shifting political and economic forces' impact on society, 'culture' has become an ever-evasive chimera defying definition. In these nebulous times, designers' familiarity with uncertainty fits comfortably with the ambiguity of play.

HK's growth has evolved from the exploitation of low-cost labor in manufacturing and shipping decades ago and, as most economists predicted, moved away from these economically unsustainable strategies to develop a service, knowledge and creative economy with world-leading communication, banking, and educational institutions, working with confidence that the rule of law and access to information are upheld.

HK's open market, intellectual property regulations, and unrestricted access to information makes good economic sense for China's future development. HK's specific cultural context is an opportunity for China: situated in a Chinese context, open to the rest of the world, with a history of global entrepreneurial cultural interchange, nurturing its design education context, places it in a privileged position to contribute to the development of the creative industries across China. HK's singular cultural context, where an international blend of influences informs design, creativity, and innovation, aptly serves as an experimental launchpad to propose solutions for Chinese education contexts and for emerging nations' innovation industries.

The open, inclusive, empathetic framework of play enriches design pedagogical framework, favoring the multicultural perspectives inherent to global design practice, while fostering the conditions necessary for the ongoing generation of cultural identities. The integrative, inclusive, ontological nature of play is a framework for constructive cultural interchange. HK's privileged - 'Special' - vanguard perspective on design, informed by Chinese heritage and a free access to international worldviews, will serve as a model for the development of emerging design practices across the region.

People are not machines: they are complex adaptive systems interacting with shifting social environments: paraphrasing Manu, the ecology of play is the ecology of possibility: to the inert nature of things, play brings life. In the age of postindustrial design and the new economics, our role as design educators is no longer to facilitate the design of interactions with products or interfaces but to facilitate design interactions with possibilities. This paper has revealed how play underlines design processes, and is instrumental in design education as a demonstrative design thinking approach, or in professional practice, as a positive agent of cultural transformation. As SD consolidates its design education, it integrates play in its curriculum to better design thinking pedagogy in HK's and China's sociocultural context. This context presents an ideal springboard for testing out homegrown methods, contextualizing regional specificities in global dynamics, for the future development of culturally viable and sustainable design education experiences.

REFERENCES

Abrami, R. M., Kirby, W. C., & McFarlan, F. W. (2014). Why China can't innovate. *Harvard Business Review*, *92*(3), 107–111.

Amabile, T. M. (1979). Effects of external evaluation on artistic creativity. *Journal of Personality and Social Psychology*, *37*(2), 221–233. doi:10.1037/0022-3514.37.2.221

Amabile, T. M. (1982). Social psychology of creativity: A consensual assessment technique. *Journal of Personality and Social Psychology*, *43*(5), 997–1013. doi:10.1037/0022-3514.43.5.997

Amabile, T. M. (1996). *Creativity in context: update to 'The social psychology of creativity*. Boulder, CO: Westview.

Atkinson, D. (1999). TESOL and Culture. *TESOL Quarterly*, *33*(4), 625–654. doi:10.2307/3587880

Baer, J. (1994). Performance assessments of creativity: Do they have long-term stability? *Roeper Review*, *17*(1), 7–11. doi:10.1080/02783199409553609

Baudrillard, J. (1994). *Simulacra and Simulation*. University of Michigan Press (first published 1981 in French).

Biggs, J. (1996). Western misperceptions of the Confucian-heritage learning culture. In D. Watkins & J. Biggs (Eds.), *The Chinese learner: Cultural, psychological, and contextual influences* (pp. 45–67). Hong Kong: CERC and ACER.

Biggs, J. (1998). Learning from the Confucian heritage: So size doesn't matter? *International Journal of Educational Research*, *29*(8), 723–738. doi:10.1016/S0883-0355(98)00060-3

Biggs, J., & Watkins, D. (1996). *The Chinese learner in retrospect*. Melbourne: CERC, Hong Kong, and ACER.

Biggs, J. B. (1989). Approaches to the enhancement of tertiary teaching. *Higher Education Research & Development*, *8*(1), 7–25. doi:10.1080/0729436890080102

Brown, T. (2008). *Tales of creativity and play*. Presentation at the Serious Play Conference. Retrieved October 31, 2015, from http://www.ted.com/talks/tim_brown_on_creativity_and_play.html

Caillois, R. (2001). *Man, play and games*. Urbana, IL: University of Illinois Press.

Clark, H. (2009). Back to the future, or forward? Hong Kong design, image, and branding. *Design Issues*, *25*(3), 11–29. doi:10.1162/desi.2009.25.3.11

Cross, N. G. (1982). Designerly ways of knowing. *Design Studies, 3*(4), 221–227. doi:10.1016/0142-694X(82)90040-0

Cross, N. G. (1990). The nature and nurture of design ability. *Design Studies, 11*(3), 127–14. doi:10.1016/0142-694X(90)90002-T

Debord, G. (1970). *The society of the spectacle*. Detroit, MI: Black & Red.

Dorst, K., & Reymen, I. (2004). Levels of Expertise in Design Education.*2nd International Engineering and Product Design Education Conference (IEPDE).*

Dunne, A. (1999). Hertzian tales: electronic products, aesthetic experience and critical design. London, UK: Royal College of Art Computer-Related Design Research Studio.

Elliot, M. (2008, January 17). A tale of three cities. *Time Magazine.*

Eno, B., & Schmidt, P. (1975). *Oblique strategies: over one hundred worthwhile dilemmas.* Retrieved November 2, 2015, from http://www.rtqe.net/ObliqueStrategies/

Fletcher, A. (2001). *The art of looking sideways*. London, UK: Phaidon.

Fry, T. (2003). The 'futurings' of Hong Kong. *Design Issues, 19*(3), 71–82. doi:10.1162/074793603768290856

Fung, A., Lo, A., & Rao, M. N. (2005). *Creative tools*. Hong Kong: The Hong Kong Polytechnic University.

Gaver, W. W., Bowers, J., & Boucher, A. (2004). The drift table: designing for ludic engagement. *Proceedings of the CHI EA '04 Conference*. ACM Press. doi:10.1145/985921.985947

Ghiselli, E. E., & Siegel, J. P. (1972). Leadership and managerial success in tall and flat organization structures. *Personnel Psychology, 25*(4), 617–624. doi:10.1111/j.1744-6570.1972.tb02304.x

Hennessey, B. A., & Amabile, T. M. (1988). The conditions of creativity. In R. Sternberg (Ed.), *The nature of creativity* (pp. 11–38). New York, NY: Cambridge University Press.

Highet, G. (1950). The art of teaching. Alfred A. Knopf, 194. In P. Rand, (1965) *Design and the play instinct*. In G. Kepes (Ed.), *Education of vision* (p. 156). New York, NY: George Braziller Inc.

HKSARG. (2011). Thematic report: ethnic minorities. In *The 2011 population census*. Census and Statistics Department, The Hong Kong Special Administrative Region Government. Retrieved November 8, 2015, from http://www.census2011.gov.hk/pdf/EM.pdf#Page=28

Huizinga, J. (1972). *Homo Ludens: A study of the play element in human culture*. Boston, MA: Beacon Press.

IDEO. (2003). *IDEO method cards: 51 ways to inspire design*. William Stout.

Keeley, L., Walters, H., Pikkel, R., & Quinn, B. (2013). *Ten types of innovation: the discipline of building breakthroughs*. New York, NY: Wiley.

Kristiansen, P., & Rasmussen, R. (2014). *Building a better business using the Lego serious play method*. New York, NY: Wiley.

Leclerc, R. (2005). 22°15'N, 114°10'E. *Urban China, 5*. (first published 1999 in Tofu Mag-Zine)

Leclerc, R., & Wan, B. (2012). Design play-interactive education toolkit for design project planning and management. In *Design Ed Asia 12 Conference Proceedings*. Hong Kong: School of Design, The Hong Kong Polytechnic University.

Louie, K. (2005). Gathering cultural knowledge. Useful or use with care? In J. Rand & J. Carroll (Eds.), *Teaching international students. Improving learning for all* (pp. 17–25). Abingdon, UK: Routledge.

Manu, A. (1998). *Tool toys: tools with an element of play*. Copenhagen, Denmark: Danish Design Centre.

Manu, A. (2007). *The imagination challenge: points of departure for strategic creativity and innovation*. New Riders. Retrieved November1, 2015, from http://english.360elib.com/datu/H/EM315566.pdf

Manzini, E. (2003). Hong Kong as 'laboratory of the future'. *Design Issues, 19*(3), 59–7. doi:10.1162/074793603768290847

Margolin, V. (2007). Design, the future, and the human spirit. *Design Issues, 23*(3), 4–15. doi:10.1162/desi.2007.23.3.4

Margolin, V. (2009). Design in history. *Design Issues, 25*(2), 94–105. doi:10.1162/desi.2009.25.2.94

Mayfield, M. (2011). Innovation. In M. A. Runco & S. R. Pritzker (Eds.), *Encyclopedia of creativity* (2nd ed., pp. 658–666). Academic Press. doi:10.1016/B978-0-12-375038-9.00122-9

McLuhan, M., & Fiore, Q. (2005). *The medium is the massage*. Berkeley, CA: Gingko Press.

Munari, B. (1971). Design as art. London, UK: Penguin Classics.

Parten, M. (1932). Social participation among preschool children. *Journal of Abnormal and Social Psychology*, *28*(3), 136–147.

Pellegrini, A. D., & Smith, P. K. (2005). *The role of play in human development*. Oxford, UK: Oxford University Press.

Piaget, J. (1959). *The language and thought of the child*. Psychology Press.

Pope Edwards, C. (2000). Children's play in cross- cultural perspective: A new look at the six cultures study. *Cultural Research*, *34*(4), 318–338. doi:10.1177/106939710003400402

Quacquarelli Symonds. (2015). *QS Best Student Cities 2015*. Retrieved November 2, 2015, from http://www.topuniversities.com/city-rankings/2015#sorting=rank+custom=rank+order=desc+search=

Rand, P. (1965). Design and the play instinct. In G. Kepes (Ed.), *Education of Vision* (pp. 156–174). New York, NY: George Braziller Inc.

Rittel, H., & Weber, M. D. (1973). Dilemmas in a general theory of planning. *Policy Sciences*, *4*(2), 155–169. doi:10.1007/BF01405730

Rousseau, J.-J. (1979). *Emile, or on education*. New York, NY: Basic Books.

Ryan, J. (2010). 'Chinese learners': misconceptions and realities. In J. Ryan & G. Slethaug (Eds.), International education and the Chinese learner (pp. 37-56). Hong Kong: HKU Press. doi:1.5790/hongkong/978988802845.003.0003

Salen, K., & Zimmerman, E. (2003). *Rules of play – game design fundamentals*. Cambridge, MA: MIT Press.

Salili, F. (1996). Accepting personal responsibility for learning. In D. Watkins & J. Biggs (Eds.), *The Chinese learner: cultural, psychological, and contextual influences* (pp. 85–105). Hong Kong: CERC & ACER.

Schiller, F. (2004). *Letters upon the aesthetic education of man, letter XV*. Dover Books.

Simon, H. (1969). *Sciences of the artificial*. Cambridge, MA: MIT Press.

Smolucha, L., & Smolucha, F. (1986). L. S. Vygotsky's theory of creative imagination. In *Proceedings of the Annual Convention of the American Psychological Association.*

Sutton-Smith, B. (1997). *The ambiguity of play.* Cambridge, MA: Harvard University Press.

Volet, S. E., & Renshaw, P. D. (1996). Chinese students at an Australian university: adaptability and continuity. In D. A. Watkins & J. B. Biggs (Eds.), *The Chinese learner: Cultural, psychological and contextual influences* (pp. 205–220). Hong Kong: HKU Press.

Von Oech, R. (1992). *Creative whack pack.* Stamford, CT: US Games Systems.

Watkins, D. A., & Biggs, J. B. (1996). *The Chinese learner: cultural, psychological and contextual influences.* Hong Kong: HKU Press.

Wong, W. (1973). *Principles of two-dimensional design.* New York, NY: Van Nostrand Reinhold.

Wong, W. (1977). *Principles of three-dimensional design.* New York, NY: Van Nostrand Reinhold.

Zimmerman, J. (2006). *Innocents abroad: American teachers in the American century.* Cambridge, MA: Harvard University Press. doi:10.4159/9780674045453

KEY TERMS AND DEFINITIONS

Creativity: A product is considered creative to the extent that it is both a novel and an appropriate, useful, correct, or valuable response to an open-ended task. Experts can recognize creativity operating in their domain when they see it (Amabile, Hennessey, 1979, 1988). Creativity usually involves a combination or manipulation of information and knowledge to generate new ideas.

Cultural Agency: The ability to foster cultural development and evolution and nurture social change.

Design: Design is the transformation of existing conditions into preferred ones (Simon, 1969). In an increasingly artificial world, design *shapes* culture.

Design Project: A privileged moment out of normal time and space akin to the 'magic circle' of play where designers and project stakeholders can experiment with 'what if?' scenarios in a trustful atmosphere.

Design Thinking: In this paper the term refers to its understanding as a cognitive process for innovation, and in conjunction with its underlying play characteristics such as open-ended explorative play, construction of knowledge and solution through quick iterations, creative play, role play and social stakeholder engagement, and divergent/free play and convergent/rule-based game play processes.

Innovation: Creativity purposefully applied to solving a given problem. This can be applying a completely new idea to the given problem or applying an existing innovation in a new setting and can be viewed as occurring on the individual, team, organizational, industry, and societal levels (Mayfield, 2011). Keeley (2014) sees 10 types of innovation classified along three categories (Configuration, Offering, Experience), and advocates for a disciplined construction of strategies articulating tactics listed under each Type.

Play: Play may be seen as a voluntary form of participation in activities that need not be prescribed or have an apparent practical purpose, and is enjoyed by adults and children alike. Play has biological (individual) and social (group) developmental functions. As such, play is also an agent of cultural evolution.

Chapter 9
The Neurolecturer as Model for Design Education:
Fostering Creativity and Innovation Based on Neuroscience

Miguel Rivas
University of Central Lancashire, UK

Giovanni J. Contreras Garcia
The Hong Kong Polytechnic University, Hong Kong

ABSTRACT

Are there lecturers capable of educating a digital generation of players in state-of-the-art innovation and creativity techniques? Far-reaching challenge to accomplish, but hard to avoid if governments and institutions want to overcome the future challenges of the global economy, particularly emerging nations such as the 'BRICS' and 'MINT.' Education design can be instrumental to support a deep transformation of society, but the expertise needed in the classroom must be updated for a new era, characterized by educational approaches aimed at developing multiple intelligences. Teaching styles that match students' personality will enrich that process by taking into account their behavior when making pedagogic decisions. Cognitive neuroscience applied to learning will be a plus to ensure a deep transformation of design education in the in 21st century. The conceptualization of a new kind of lecturer, capable of carrying out this application is the center of this chapter.

DOI: 10.4018/978-1-5225-0911-0.ch009

INTRODUCTION

According to The Economist (2013), and American business 'think tanks' (Shierholz, 2014), the latest crisis has three main causes: a lack of skilled workers, a lack of leadership abilities from those who caused it, and a lack of innovative creativity to anticipate it. An opportunity that cannot be wasted to nurture men and women socially committed through education.

A new lecturer in Design Education should emerge to move emerging nations towards a focus on creativity. The GDPis not exclusively based on economic indicators anymore (Bauman, 2001), but on intangibles like welfare, health and education.

Design educators are key players in spreading creativity and innovation in emerging nations by teaching differently. That will be easier with state-of-the-art syllabus characterized by passion and authenticity applied to Design practice.

Those who currently teach and aspire to upgrade their performance already know that emotions are rooted physically, and profoundly influence not only what people reason and feel, but how they reason and feel. If Design students do not know that, either they will not be able to make proper decisions, or will make self-defeating ones as future leaders.

Two main questions are considered in this chapter:

1. Are educators/institutions in China and other emerging nations prepared for a post-crisis time, characterized by the need for a more creative and emotional approach, capable of creating more suitable educational environments?
2. What kind of lecturer is the answer? What kind of lecturer can develop these environments and upgrade performance in both Design education and creativity-related industries? Furthermore, how is this lecturer going to be trained?

A designer is a holistic individual, used to go beyond limits, with an 'out of the box' mindset as driver to learn on permanent basis. A designer is someone who loves disruptive innovation, over 'sustainable' or incremental one (Christensen, 2011). Only Individuals alike can are capable of training them.

Descartes was wrong: "Cogito ergo sum" is not valid anymore (Damasio, 2005), a single rational way of education is not acceptable when emotions are taken as important as technical knowledge. Reason and emotion are both essential to form a holistic professional; emotions are innate and technical knowledge comes from university, it is necessary to get both in the right balance. Graduation from the 'school of life' is as important as graduating from University for any designer interested in social trends, season styles and local fashion. But are design lecturers in emerging nations transmitting this message?

EMOTIONAL EDUCATION AND CREATIVITY: A CASE FROM SPAIN

The Government of The Canary Islands in Spain, made a challenging decision in early 2014 by approving the inclusion of 'Emotional Education and Creativity,' in the primary education curriculum as a compulsory subject. The change aims to contribute on the recognition and expression of emotions, regulating, controlling and using them effectively. According to the technical report that recommends its inclusion, emotions are educable and, and this is something schools cannot ignore.

This initiative criticizes the accent set on classical education in the development of intellectual abilities to the detriment of the emotional ones, advocating the creation of spaces and specific times for the new learning.

This essential subject, includes three thematic blocks to be developed year by year throughout the six-year of primary school:

1. **'**Awareness and emotional literacy**'**, in which students learn to perceive, validate, accept, classify and communicate one´s emotions and those of others.
2. **'**Emotional regulation**'**, where they learn to manage conflicts and modify emotions appropriately, according to contexts and relationships.
3. **'**Creativity**'**, where they develop self-confidence on their own creative abilities, as well as a favorable attitude towards reality and novelty. Knowing how to recognize and handle fear and frustration, helps them to develop resilience and to control their impulses (Infocop, 2014).

THE NEUROLECTURER

Emerging nations should consider developing similar models, if they aim to succeed in the global competition for creativity and innovation.

Creative industries are key for emerging nations like China, because as their economies grow, these countries must find other competitive advantages beyond low-cost manufacturing. However, creative industries require qualified educators who can meet the expectations of students and industry. China has nowadays one of the largest number of institutions delivering some type of design degree in the world (Siu & Contreras, 2016), yet, many have highlighted important issues with the Chinese education system that hinder the education of creative professionals (Wertime, 2015).

Former CEO of Hewlett-Packard and US presidential candidate, Carly Fiorina's comment that Chinese "are not terribly imaginative" (Wertime, 2015), has been criticized as a sweeping judgement, but it highlights a common perception; that the world's second largest economy is short on home-grown innovation, as it is of the business and academic cultures necessary to nurture it. Chinese education has been widely criticized for having an inhibiting effect upon creativity, and for being more suitable for the formation of engineers rather than scientists (Allen-Ebrahimian, 2015). However, it is not for lack of efforts; the Chinese government has poured billions of dollars into research, development and innovation, and may surpass the corresponding spending in the USA by 2019.

Similarly, President Xi Jinping has emphasized innovation in his speeches; for the past 4 years, China has filled more patent applications than any other country, although state news agency Xinhua has described the quality of those patents as "poor' (Allen-Ebrahimian, 2015).

While some universities such as the Hong Kong Polytechnic, are currently combining innovation, design, and business. It is important to remember that Professor Mintzberg's claimed that MBAs should move from the current focus on Management 'by analysis,' to Management 'by attitude' (Mintzberg, 2001). Professor James Heskett from the Harvard Business School, also explores the emerging world a new manager that he calls 'neuro-manager,' someone who can learn which part of the brain governs how we feel, how we respond to stimuli, and how we react to challenges (Heskett, 2014). When educators learn about how the brain processes, recognizes, remembers, and transfers information at the level of neural circuits, they manifest that knowledge in the classroom, and that empowers students. This, as will be seen later, will constitute the foundations to determine the competences of the new type of lecturer that is being conceptualized.

This chapter aims, through a meta-analysis based research, and in-depth interviews, to conceptualize a new type of lecturer, namely a *'neurolecturer,'* as well as to describe a personality typology to detect 'neuro-teaching talent' in early stages. In addition, it attempts to determine the competences of this new kind of teaching professional, anticipating that, as education transitions from an 18th century model based on standards, towards a twenty-first century model based on individual needs, lecturers knowledgeable of the role of multiple-intelligences, and the role of emotions in learning, will be the norm.

DEFINING THE NEUROLECTURER

What will this new sort of lecturer be like when creativity and innovation are a must? This new professional will be the result of not only a professional, but a 'personal transformation age' related to reasoning and emotion, which is to take over the 'information age' (Rivas et al, 2012). A neurolecturer is a holistic person, with a deeper understanding of the brain's cognitive and emotional functions, thus the name. This person embraces the way people—students—think, feel and act, to understand the way they learn as a whole. This is the tremendous opportunity we have these days, to take advantage of a better knowledge of how our brain works, to positively influence educational practice. These lecturers rely on their knowledge of neurology, to strengthen their role as teacher, and to identify poor teaching habits in classrooms. At the same time, students welcome tutorials that take into account student's emotional skills, and which lead to an enriched debate. The neurolecturer does not care about performance metrics, but on his own commitment, self-education and capacity to develop personal skills and abilities. These individuals know that the notion that IQ should be a benchmark of an individual's intellect should be dismissed (Rivas, 2012).

In training future designers; understanding how people feel is as important as understanding how they reason. Syllabus' related to creative disciplines can consider this to add value to a 21st century economy characterized by services and a more humanistic orientation. Knowledge of oneself and teaching vocation are basic to a neurolecturer. These kind of lecturers must be aware their own knowledge limitations; first of all about what they do not know, what they know partially, and what they know 'transversally' (Rivas et al, 2012). This person, eager to get a holistic approach, should be aware of how crucial it is to play a revolutionary role.

Any neuro-oriented professional, holds a balanced emotional-rational cognitive mix, characterized by:

1. The determination to discover passion in something concrete;
2. The compromise of making that passion valuable, by making it unique;
3. Feeling one's way, ignoring the risks before making a deal;
4. Listening instead of hearing, even in periods of silence or unexpected messages;
5. Observing and not only seeing, generating behaviors to get empathy;
6. Sniffing the air as a prior step to sound out the environment; and
7. Sampling and tasting to make sure fair decisions (Rivas et al, 2012).

These individuals have a deep understanding of the value of culture, regardless of which one. They look for an interactive communication, and not only the transmission of information. They search for experience and emotions, and not just for knowledge. Holistic designers like holistic people are used to learning everything, and not only the study of specific matters. They take results as the consequence of, and not only an exclusive goal. They grow up through learning from human relationships and love people promotion and long-life learning. They make daily decisions and prefer divergent thinking and discussing against a simply verbalization of orders.

The purpose of a neurolecturer as future design educator, is to lend education institutions a new focus; turning towards the neurological and emotional fields, and in doing so, searching for new tools that could deal with what is unclear. Holistic designers discover the roots of what they and their team think, say, feel and do. All of this brings about a scenario of unprecedented learning improvements and continual growth never experienced before, and so expeditious, so competitive, that imitations will be hard to make.

For the neurolecturer, a great source of knowledge is one's own experience, one that does not need to be memorized, does not demand previous training and does not need to be analyzed once reached will sprout. It only needs to be enjoyed in person, as the most seductive act, taking advantage of any problem or adversity as the perfect opportunity to learn (Rivas et al, 2012).

These unusual individuals will never remain prisoners of the place in which they were born. They won't even speak their mother tongue in detriment of another, or be part of a culture, social or financial group that doesn't allow them to distance themselves from the events and situations in which they are immersed, so as not to lose a sense of perspective. And all the aforementioned can be justified by the decision taken by two adults (their progenitors) who decided to have a baby without musing on the role or kind of work that the child could do later in life. Unaware if this role would consist of being an employee of a company in the industrial sector or in services or in carrying out a particular job as a self-employed person.

AN AGENT OF DISRUPTION

Recent research (Doin, 2012) has confirmed what has been for long suspected, that across all levels of the education system, institutions still include in their make-up and organization a configuration similar to that established at the forefront of the Industrial Revolution of the 18th century. This productive and economic model still exists in many nations today. It is from this model, that a standard syllabus producing uniform individuals—working class—emerged. It is also from this model, that technical and functional subjects were given priority over artistic-oriented ones, as technical subjects are easy to measure—thus evaluate (Robinson, 2010).

Some of the features that define this educational system can easily be observed in many schools and colleges:

1. Bells and sirens which regulate the movement of pupils, inappropriate in the society of TIC[4].

2. Separated by departments or enclosed areas, as if avoiding individual differentiation.
3. Evaluation of memorization or deductive reasoning without considering emotional issues.
4. Graduation at the end of a cycle, as if referring to the fabrication date.

The situation gets worse with the hindrance of the Enlightenment with its 'encyclopedism' and the reduction of intelligence to mere deductive reasoning, not the ideal model to educate designers. Such a model, discriminates against brilliant people for not fitting in the system, and encourages the accumulation of knowledge as if the mind was a warehouse, whose value depends on the number of articles it can store and remember, what Noam Chomsky refers to as 'the empty vase' model of education, in which the immersion in the cognitive experience, the preparation for an early assumption of risks, and with the promotion of creativity and innovation are ignored. The composition of standard courses based on age, without personal information about each student, does not facilitate awareness of the talent destiny of each student. Such a system, promotes knowledge which is harmful, slow and evasive.

NEUROLECTURER: COMPETENCES FOR DESIGN EDUCATION

No person who is not a great sculptor or painter can be an architect. If he is not a sculptor or painter, he can only be a builder. (John Ruskin)

According previous research (Rivas, 2012) there will be a three-fold self-development process for any neurolecturer. This process derives from extrapolating the work of Heskett (2014), who conceptualizes a 'neuromanager,' and which is adapted here to the classroom environment:

1. Searching, for what is not under control because of the opportunity to gain vivid experiences.
2. Confirming what is perceived through the senses, and intuited, as a way of emotional development.
3. Contrasting what is known, with what is thought, and gain new knowledge from this comparison.

Anyone who aspires to be a neurolecturer faces the challenge of being convinced that the first vocational symptom of this new role as self-manager and neurologist, is to give up, without complaint, any comfort zone; these individuals are never prisoners of their origin whether in terms of language or geographic location. Similarly, a neurolecturer must be aware of the level of mastery that they have of any subject they aspire to teach:

1. **Unknown**: From general to specific;
2. **Partially Known**: Lack of practice or dated knowledge; and
3. **Transversally Unknown**: Matters beyond the core-subject.

Remaining a prisoner of one's place of birth is not inspirational for a designer, willing to gain knowledge of overseas trends and avant-garde movements, but what is the message that design students in China are receiving from their lecturers? According to personal experiences, old habits and established routines remain against new trends and overseas influence in China. It is therefore important to renew the education system as soon as possible in order to facilitate Design education performance on creativity and innovation.

In such a scenario, neurolecturers assume that a university degree or the lack of one, is nothing but a starting point, and never a destiny. They also understand that practical knowledge is acquired better when senses that intervene in the learning process are engaged, a "cognitive-sensorial immersion" that occurs when faced with the unknown or "transversally unknown" (Eker, 2007).

For a neurolecturer, the classroom is a tool for getting to know people who can become allies in the development of pedagogical opportunities (Rivas, 2012).

The communicative capacity of these individuals, should be such, that they make their presence felt, without necessarily speaking a word. They are familiar with the body language of the cultures that they work with, are capable of approaching the language of music, theatre, or other artistic form. In that way, they gain in sensitivity.

The main personality traits of the neurolecturer are:

- Empathetic
- Holistic
- Creative
- Disruptive
- Open-minded
- Sensitive
- Entrepreneurial
- Ambitious
- Curious
- Responsible
- Determined

How many of these attributes are the attributes of the average lecturer in Design schools? How about Design schools in China?

RICHARD BRANSON: A CASE STUDY

Knowledge is not what you remember but what you don't forget. (Confucius)

The individual who takes the plunge into neurolecturing must be interested in finding the greatest opportunities, without being limited by conditions (Rivas, 2012). This is because they do not exist in the mind, so these individuals can identify and enhance the most unimaginable possibilities.

Sir Richard Branson demonstrated it himself, as a holistic person and service designer, he has become the prototype of an innovative and disruptive neuroentrepreneur. Founder of Virgin, the company that revolutionized service design by going from music delivery to fitness gyms to suborbital trips, Branson was not inhibited by the fact that, in another, not so distant epoch, the space travel sector had been reserved, almost exclusively, to astronauts. It wasn't hard for him to change this paradigm and launched 'Virgin Galactic' a pioneer company in space travel, to cover an emotional necessity.

In a great example of what a neurolecturer could be, in 2007, he talked about the first suborbital trip teaching to the global audience about how to make it possible. Such individuals never lose the opportunity of participating in what seems remote, bizarre and worth diving for. So long, entering those hidden zones they find personal development facilitated. They will never be unwilling to use all their physical, mental and intellectual strength if, in return, they gain in knowledge or experience. What does this mean in the case of an educator?

NEUROLECTURER: THE SCIENCE BEHIND

The brain is a product of millions of years of evolution, and is engaged in the fast hunting and capturing of rewards, aligning as many neurotransmitters as possible for this purpose, and in some cases triggering off aversion to the risk or pain. The functions that the brain has evolved to perform are, amongst other:

1. To live in society, get known and form relationships;
2. To develop activities towards concrete and specific goals;
3. To avoid danger or the aversion at loss or the negative;

4. To interpret data in order to make judgements and behave oneself; and
5. To make decisions and assume the consequences.

These influences all lecturers on their decision making process, the big difference is that neurolecturers are aware of it. And thus are fitter to lead a paradigm shift that would favor the development of creativity and innovation (Rivas, 2012). In consequence, a core syllabus should be defined to transform lecturers into neurolecturers, if emerging nations are determined to foster creativity and innovation through their education system.

Recently, Velasco and Megia (2011) have confirmed that the brain does not seem to be very well-prepared for dealing with the complexities of modern life, and our education system is one of such complexities. The current make-up of the brain has the imprint of its evolutionary past with three areas going from the most ancient to the most recent; these are described according to their phylogenetic chronology:

1. **The Reptilian Brain:** This regulates the basic body functions such as cardiac rhythm, breathing and temperature control; mostly subconscious.
2. **The Limbic System:** This controls the subconscious emotional responses such as the state of fear. It is a source of motivations and primal emotions which includes euphoria and panic.
3. **The Prefrontal and Cerebral Cortex:** The last structure to be developed phylogenetically. Within it we can find essential functions such as logical reasoning, abstract thought, calculus, learning and decision-making (Mac Lean, 1990).

It is extremely important that the neurolecturer's mental activity obtains the greatest 'reward' with the least amount of pain or 'loss.' The ultimate goal of a neurolecturer will be to use his knowledge of the brain functions to deliver unique and rich lessons, rationally and emotionally equilibrated.

FUTURE RESEARCH

Adversity is the cornerstone and inspirational source that drives evolution (Darwin, 1859). The Homo-Sapiens defeated The Neanderthal by discovering and perfecting his shooting technique, and replacing the cave as living environment, leaving a cannibalistic way of life behind (Rivas et al, 2012). In a similar way, a new species, namely "Neuro-Sapiens" is being proposed as a new evolutionary step, in which neurolecturers aware of the importance of keeping a balance between reasoning and emotions are due to run the education system.

At this stage, twenty first century Designers could play a key role as early "neuro-sapiens," moving forward design in emerging nations. But it is necessary to develop the necessary theories that support both: their emergence, and their work, theories of intelligence show how to train this 'new species,' at the moment, the closes that we have, are Gardner's (1993) and Goleman's (2005) theories. If adversity is an inspirational source to be more creative, it has not yet yielded substantial results in Asian countries, some of which have gone through difficult historical periods; Chinese education is considered more suitable for cultivating engineers than artists (Weir, 2015). Finding the answer to this seemingly paradox undoubtedly demands future work of research and analysis.

CONCLUSION

Culture and traditions are among the things that generate habits and routines which attempt largely against renewed ways of doing things; neither teachers nor educational institutions from China and emerging countries are ready to face a post-crisis time in which a more creative consciousness will be essential to survive amidst increasing competition. Educational governance is a tough issue but clear and informed decisions are a must. They should be made on time to start the process of transforming the current education paradigm based on developing student's rational abilities.

The brain houses the mind and education develops it. Identifying educational goals and challenges will be the kind of mind that will characterize future generations of design lecturers, making decisions that will set the path that, nations, corporations, and education institutions shall follow. In this chapter, a new concept of lecturer has been presented, an education professional with a holistic approach, and knowledgeable about brain processes and the role of emotions in a person's development.

The neurolecturer possesses necessary cognitive and emotional skills to be the classroom's architect and manager. A great opportunity for visionary countries, but also for those lecturers eager to be part of this paradigm shift, never before has a lecturer faced his profession like an emotional immersion. Just like a mental athlete, the neurolecturer shall train for the big race. The decision of whether to pursue that challenge or not, will depend not so much on his natural skills, but on a commitment to enhance them, and develop the right abilities to achieve a higher performance in the benefit of his/her addressees.

Depending on their ability to crystalize their creative potential, emerging nations, representing the hope of growth in the world, will succeed in the global economy. A new approach that enhances syllabus in design schools, based on this orientation, can start an educational revolution with the neurolecturer at the center.

ACKNOWLEDGMENT

Special appreciation and gratitude to Sofia Odiaga for the initial Spanish to English translation, and to Giovanni J. Contreras for the subsequent extensive revision and adaptation.

REFERENCES

Alonso García, C. (2008). Learning Styles: Present and Future. *Journal of Learning Styles, 1*(1). Retrieved on 25 April, 2016 on: http://learningstyles.uvu.edu/index.php/jls/article/view/73

Anonymous. (Essay), translation by Allen-Ebrahimian, B. (2015). *Why do Chinese lack creativity?* Retrieved from http://foreignpolicy.com/ 2015/06/23/ china-innovation-creativity-research-patents/

Barboza, D. (2006). *China Says One of Its Scientists Faked Computer Chip Research.* Retrieved from http://www.nytimes.com/2006/05/14/technology/ 14chip.html?_r=2&

Bauman, Z. (2008). *The Art of Life*. Polity Editions.

Bermejo Velasco, P. E., & Izquierdo Megía, R. J. (2011). *Cerebro rico, cerebro pobre (Brainrich, Brainpoor)*. Cultiva Libros.

Caldwell-Harris, C., & Aycicegui-Dinn, A. (2008). *Emotion and lying in a non-native language*. Elsevier.

Christensen, C. (2011). *The Innovator's Dilemma: The Revolutionary Book That Will Change the Way You Do Business*. Harper Business. doi:10.15358/9783800642816

Contreras, G., & Siu, K. (2016). *Design Education in China: Fostering creating and innovation in China and emerging nations*. Retrieved from: http://gicoga.wix.com/design-edu-in-china#!introduction/cee5

Damasio, A. (2005). *Descartes Error: Emotion, Reason and the Human Brain.* Penguin Books.

Darwin, C. (1859) The origin of species. Wordsworth.

Doin, G. (Director). (2012). *The Forbidden Education* [Motion Picture].

Eker, T. H. (2007). *Secrets of the Millionaire Mind: Think rich to get rich.* Piatkus.

Gardner, H. (1993). Multiple Intelligences: The theory in practice. Basic Books.

Goldberg, E. (2001). *The Executive Brain*. Oxford University Press.

Goleman, D. (2005). *Emotional Intelligence: Why it can matter more than IQ*. Bantam Books.

Grignon, P. (2012). *Money as Debt*. Fábula de Goldsmith. Retrieved from http://. asisonlascosas/el-dinero-es-deuda.

Herrmann, N. (1989). The Creative Mind. Brain Books.

Heskett, J. (2014). *Are we entering an era of neuromanagement?* Retrieved from http://hbswk.hbs.edu/item/are-we-entering-an-era-of-neuromanagement

Hudson, A. (2014). *Howard Gardner's Learning Styles*. Retrieved from: http://www. ehowenespanol.com/estilos-aprendizaje-howard-gardner-sobre_327898/

Kahneman, D., Tversky, A., & Slovic, P. (1982). *Judgement under uncertainty: Heuristic and biases*. Cambridge University Press. doi:10.1017/CBO9780511809477

Lane, C. (2012). *Multiple Intelligences*. Retrieved on 29 November, 2015 from: http://www.tecweb.org/styles/ gardner.html

Mac Lean, P. (1990). *The Triune Brain in Evolution: Role in Paleocerebral functions*. Springer.

Mintzberg, H. (2004). *Managers, not MBA's. A critical vision of company management and management training*. San Francisco, CA: Berret-Koehler.

Rivas, M. et al.. (2012). *Tu Cerebro lo es todo (Yourbrainiseverything)*. Plataforma Editorial.

Robinson, K. (2010). *Bring on the Learning Revolution*. Retrieved from http//ted. com/tals/lang/eng/sir_ken_robinson_bring_on_the_revolution.html

Salas Silva, R. (2008). *Estilos de Aprendizaje a la luz de la Neurociencia (Learning Styles through Neuroscience)*. Bogotá, Colombia: Cooperativa Editorial Magisterio.

Sharp, D. (1987). *Studies in Jungian Psychology by Jungian Analysts*. Inner City Books.

Shierholz, H. (2014). *¿Is there really a shortage of skilled workers?* Economic Policy Institute. Retrieved from http://www.epi.org/publication/shortage-skilled-workers/

The Economist. (2013). *Crash Course*. Retrieved on 24 November, 2015 from: http://www.economist.com/news/schoolsbrief/21584534-effects-financial-crisis-are-still-being-felt-five-years-article

The New York Times. (1989). *What's in a name?* Retrieved from http://www.nytimes.com/1989/04/02/magazine/what-s-in-a-name.html?pagewanted=all

Vázquez, J. (2016). *Cincotópicossobreinnovacióndocente (Five topics about teaching innovation).* Retrieved from http://www.universidadsi.es/cinco-topicos-innovacion-docente/

Wertime, D. (2015). *Chinese netizens to Fiorina: You're right, We don't innovate: Many in China have been complaining about their education system for years.* Retrieved on 7th May, 2016: http://foreignpolicy.com/2015/06/01/chinese-netizens-agree-fiorina-innovation-comment-education-system/

Willis, J. (2012). *A Neurologist Makes the Case for Teaching Teachers About the Brain.* Retrieved from http://www.edutopia.org/blog/neuroscience-higher-ed-judy-willis

KEY TERMS AND DEFINITIONS

Brain: The organ of the body in the head that controls functions, movements, sensations and thoughts (Merriam-Webster).

BRICS: It´s the acronym for an association of five major emerging national economies: Brazil, Russia, India, China and South Africa. The grouping was originally known as "BRIC" before the inclusion of South Africa in 2010 and was coined by Goldman & Sachs (Wikipedia).

Cognitive Neuroscience: Academic field concerned with the scientific study of biological substrates underlying cognition, with a specific focus on the neural substrates of mental processes (Wikipedia).

Disruptive Innovation or Technology: Innovation that creates a new market and value network and eventually disrupts an existing market and value network, displacing established market leaders and alliances. Also implies significant societal impact (Wikipedia).

Experience: A repertory of cognitive skills -abilities and capacities for recognising patterns- developed throughout life which enable us to approach and interpret situations with familiarity creating behavior routines (Rivas et al, 2012).

Guanxi: A Chinese social concept based on the exchange of favours, in in which personal relationships are considered more important than laws and written agreements (Rivas et al, 2012).

Holístic: Relating to or concerned with wholes or with complete systems rather than with the analysis of, treatment of, or dissection into part(Merriam-Webster).

Mind (Alternative Use for Brain): Part of a person that thinks, reasons, feels and remembers(Merriam-Webster).

MINTs: It's an acronym referring to the economies of Mexico, Indonesia, Nigeria, and Turkey. The term was originally coined by Fidelity Investments, a Boston-based asset management firm, and was popularized by Jim O'Neill of Goldman Sachs, who had previously created the term BRIC.

ENDNOTES

1 **GDP:** Gross Domestic Product.

2 **IQ**: Intellectual Quotient.

3 **GPS**: Global Positioning System.

4 **TIC**: Technologies of Information and Communication.

Chapter 10
20 Years of Doctoral Research Studies in Design in Hong Kong

Kin Wai Michael Siu
The Hong Kong Polytechnic University, Hong Kong

Giovanni J. Contreras Garcia
The Hong Kong Polytechnic University, Hong Kong

ABSTRACT

For the past 50 years, the School of Design of The Hong Kong Polytechnic University has been a pioneer in nurturing design education in Asia. It is also one of the first schools in the region to grant PhDs in "Design" serving as a role model for other universities in the region. Since the 1990s, the school has produced tens of PhD degree holders across different design specialties. However, as times change and knowledge plays a more fundamental role in the economy, the ability to produce quality design researchers, capable of creating deep and specific insights will become fundamental for the future of design in the region. In this chapter we review the emergence, growth and future of the PhD programme at the school of design of the Hong Kong Polytechnic University against a landscape of challenges of doctoral research education, particularly in the case of design. We also discuss the strategies used in the past, present and future, as the school moves into a new era of economic and geopolitical scenarios in the 21 century.

DOI: 10.4018/978-1-5225-0911-0.ch010

INTRODUCTION

Since its establishment, the School of Design of The Hong Kong Polytechnic University has been a pioneer in design education in Asia and a cornerstone of higher design education locally. It is also the 'Alma Mater' of renowned scholars like Wucius Wong and acclaimed designers like Viviene Tam. The history of the school traces back to 1964 when it was founded as part of the Hong Kong Technical College and has been a member of the International Council of Societies of Industrial Design (ICSID) for more than 30 years. During its 50 years of existence, it has consistently expanded and consolidated its educational offer in different areas of design. The school has grown and diversified its educational offer moving from the initial general certificates in design, to 2-year diplomas, to higher-diplomas, bachelor's degrees, master's degrees and finally PhDs.

Today, the school has more than one thousand registered students and an academic staff of more than 50 members from 15 different countries (De Bont, 2015). With the establishment of interdisciplinary, as well as transdiciplinary, postgraduate degrees like the 'MBA in Innovation and Design Management,'—established in conjunction with the graduate school of business— and the 'Executive Master in Meaningful Innovation,' the school has consolidated itself as a world-class design education institution as it steps into the 21st century. Several of its programmes have been ranked amongst the best in Asia and internationally and in 2014 it was ranked 1st in Asia by Quacquarelli Symonds (School of Design, The Hong Kong Polytechnic University, 2014).

In recognition to its trajectory, the school received a donation from the Hong Kong Jockey Club in 2011 to develop an entire new space to house its facilities. Since October 2013, the school is located in the Jockey Club Innovation Tower, a state of the art building designed by renowned late architect Zaha Hadid which also houses the Jockey Club Design Institute for Social Innovation (Law, 2011; De Bont, 2015).

BACKGROUND

Origins of the PhD programme at the school of design can be traced back to the early 1990s, however full-time PhD students only appeared until 1998, ever since, the number of PhD students has grown steadily. Initially, research conducted at the school tended to focus on Design Technologies, but this focus has been expanding steadily. Until recently, PhD studies at the school have happened within the different laboratories that integrate the school: Asian Lifestyle Design Lab, Asian Ergonomics Design Lab, Creativity and Design Education Lab, Digital Entertainment Lab, Information Design Lab, Interaction Design Lab, Public Design Lab, and Urban

Environments Lab. In addition, there are several other research projects under the Jockey Club Design Institute for Social Innovation (School of Design, Hong Kong Polytechnic University, 2015; De Bont, 2015). However, during the last PhD forum held and Hunan University, Professor Ilpo Koskinen announced that this strategy is changing, and in the future it is expected that most research at the school of design will fall into three main streams: Design Making, Design and Business, and Social Design (Hunnan University, 2016).

Today the PhD programme is one of the largest programmes in Asia with 45 registered students working on areas as varied as sustainable design, participatory design, public design, design and technology, design and business, and design education, and produces around 6 and 8 graduates a year (De Bont, 2015). Most of these students come to school under PhD fellowships that provide financial support for up to 3 years, however as it will be discussed later, it is common that students see themselves in the need to extend their studies beyond their 'normal study period'. The programme has and continues to be an inspiration and role model for other Universities in the region, and a springboard for students from the mainland to go abroad. As we advance further into the 21st century, the school continues to take steps towards improving the quality of its research and PhD programme, the school has recently reviewed its leadership structure, and appointed a vice-dean for research.

The programme is supported by an international and multidisciplinary team of 18 faculties who act as supervisors, co-supervisors, lecturers and examiners, and it is guided by the school's core values of: Developing an Open Mind Frame, Steering Positive Change, Establishing Visionary Entrepreneurship, Practicing Authentic Expression, and Designing Responsibly (School of Design, The Hong Kong Polytechnic University, 2015). As stated by professor Ming Xi Tang, the programme aims at attaining five outcomes with the students:

- Demonstrate a high level of understanding of issues and discourses in design theory.
- Display original and independent innovations in one of the core specialist design disciplines of the School of Design with an excellent command of knowledge relevant to the local industry.
- Be able to structure, carry out, and defend a complex and extended course of design research with; scientific, social, cultural or technological significance.
- Produce a piece of writing at a level and of a character appropriate to international publication in scholarly journals and/or other competitive venues of dissemination.
- Develop a character in solving problems with an artistic and designer's manner signifying design thinking in interdisciplinary collaborations and integration.

20 YEARS OF PhD STUDIES IN DESIGN

In this chapter, the past, present and future of the PhD programme at the School of Design of The Hong Kong Polytechnic University is reviewed against the backdrop of the complexities and challenges that surround design research and doctoral design education. This chapter discusses how those have affected the PhD programme at the school and the different ways in which the school has attempted to address them.

To accomplish this, we have conducted an archival review of information about the PhD programme and a literature review pertaining doctoral education, particularly in design. We have also interviewed a number of senior faculty and staff as well as a number of PhD students currently enrolled in the programme. Most of the information collected was gathered by means of questionnaires consisting of open-ended questions sent by email. Participants were also encouraged to answer the questionnaire in person if preferred, in which case it was treated more like a semi-structured interview.

Overcoming the Challenges of a New Field

Both, Margolin (2010) and Friedman (2015), have pointed out that design research and doctoral studies in design present a particular set of problems that sets them apart from other disciplines (Margolin, 2010) (Friedman, 2015). These issues, in part, stem from the fact that design in general and design research in particular, are relatively new fields. Yeo (2012) also points out that the fact that a good number of design researchers in the past have attempted to come up with 'their own methods' has led to 'considerable confusion' (Yeo, 2012, p. 309).

One of the problems is that the history of design studies is quite short, what we know today as Industrial and Graphic Design traces back to the 1920's and 30's, while postgraduate studies in the field only started in the mid 1970's (Friedman, 2015). In consequence, in design research we have a problem that other disciplines do not have; we have a 'meta' problem, that is, that we do not really know what needs to be researched. And while a good way to determine a good research agenda would be to look at actual design research cases from the private sector, as Margolin (2010) points out, there are a number of issues that make this difficult (Margolin, 2010).

Another problem of design research is perhaps a problem inherited from the social sciences; Margolin (2010) has done a comprehensive effort to explain some of the fundamental challenges of doing research in design. To start with, design, being part of the social world more than of the natural, entails a much higher degree of constructivism; there is no point in history where humans got the social world—including the world of design—as a pre-existing phenomenon that could be studied (Margolin, 2010). The history of human body research for example, has resulted in the creation of a community of researchers working within a set of problems relatively well defined, and whose research is sustained by a well-established pedagogic system, publications and congresses.

In contrast, design is a contingent practice, constantly 'in the making.' We have not gotten to the point where, as in the case of human body research, financial support can be made with the security that well-spent money will help to eliminate diseases. Consequently; the objective of a doctorate in design has never been properly formulated (Margolin, 2010) (Durling, 2002), and often the purpose has become more symbolic than pragmatic, so has been the need to do research, often more due to the need to sustain a title than to the desire to solve a problem.

Design research agendas for PhD programmes should be determined as problematic areas are clearly identified, which would also guarantee that graduates fit in a market that needs them. However, because the very objective of a PhD in design has never been formulated, a 'cacophony' of PhD programmes has emerged in the past. Friedman (2015) reiterates this, adding that there is also great confusion about the different kinds of PhD degrees that schools grant:

Practice-based PhD that functions within the framework of the traditional PhD, new forms of PhD, the studio PhD awarded for studio practice in the creative and performing arts and design, and degrees labeled as a practice-based PhD that are neither a studio PhD nor a traditional PhD (Friedman, 2015, p. 3).

As stated during the latest PhD Winter seminar in 2015, the PhD education offered at the school aims to be an 'outcome-based' education approach. This approach "tries to reduce the complexity in solving ill structured problems of doing a PhD in Design by enhancing the ability of the students in studying literature and gaining practical skills in design related activities like teaching" (Tang, 2016).

The degree offered at the school would clearly fall into the 'traditional' category suggested earlier by Friedman; students go through all the formalities of passing candidacy—confirmation of studies—submitting a thesis report, defending it, and following all and the same protocols of other departments of the University.

Working Out the Basics

In order to solve some of the problems that affect all Design PhD programmes, further work is necessary to define the core foundations of the PhD in Design in general. As an example, a basic literature common to all PhDs in Design has not even been worked-out, although Ken Friedman (2015) may have started an effort in this direction; the same will have to happen at institutional level as each institution determine their specific research focus (Margolin, 2010). It is also necessary to establish a clear difference between design research and design studies. Both could lead to a doctoral degree in design but with a different focus; while the second could devote their efforts to a wide spectrum of possibilities and acquiring a better understanding of design as a phenomenon, the first could aim at transforming the discipline (Friedman, 2015).

Research at the school happens within the different laboratories mentioned earlier. This means that each laboratory should—at least in theory—already have identified a set of problems to focus their research on. This does not mean that this is an issue that the school has not struggled with; when the programme started, the motivation and objective to start the PhD programme were simply that at some point 'it was natural,' and the resources were available (Tang, Interview, 2015). This was echoed by professor Ilpo Koskinen, Chair of the School Research Committee, who reckons that, in contrast to the public sector which is 'never unambiguous,' programmes like the PhD programme at the School of Design usually respond to many issues, concerns, policies, and problems (Koskinen, 2015).

Overcoming the Challenges of Sudden Growth

Why having PhDs in a field where it is not even clear what the research problems are? Any answer to this question will have to look at the shift towards research at many universities and the implications this conveys. In this respect Friedman (2015) notices that; "research involves power, politics, and wealth. Research means access to money, staff positions, and resources. For this reason, many people who do not research and want no real engagement in research, still want to designate their activities as research" (Friedman, 2015; Rossi, 2014). This trend is acknowledged at the school, where research plays a major role in faculty revisions (School of Design, The Hong Kong Polytechnic University, 2015).

This shift towards research has resulted in an explosion of PhD students in recent years (Yeo, 2012). In some cases like in China, it has been dramatic; a 10-fold increase from 5 thousand in 1990 to more than 50 thousand in 2010. This explosive growth not only has come at a price in quality as Cyranoski et al. (2011) recognize, but could also represent a major employability challenge in case of an economic 'hard landing' (Abrami, Kirby, & Warren, 2014; Cyranoski, 2011).

Closing the Knowledge Gap

Because research in design is a relatively new phenomenon, despite the latest surge in PhD programmes and student numbers, degree holders are still scarce. Therefore this growth has increased the ratio of students versus qualified supervisors (Friedman,

2015; Cyranoski, 2011). Moreover, many of those who hold a PhD were themselves supervised by people who did not have a PhD. This is because, while 'many universities appointed professors of practice who did no research,' in recent years, these universities have asked the same professors to supervise research students, which does not help to close the knowledge gap (Friedman, 2015).

In contrast to the sciences, where much of the work is done in teams, in a lab, and under constant supervision of senior researchers, in the arts and the humanities, students do not have a master-apprentice relationship to make up for the gaps in knowledge. This situation makes that knowledge gap difficult to close because to start with, both, students and their supervisors are unawareness about its existence and the need to develop their research skills.

In the school all faculty in charge of supervision are PhD degree holders, however it is plausible to think that professors who did not have a PhD themselves but with rich research experience may have supervised some of them.

Quantity vs. Quality: Meeting Demand and the 3-Year PhD

One of the greatest problems derived from this sudden growth, is the challenge it poses to maintain quality. Perhaps due the need to meet demand, many PhD programmes in Asia have adopted a three-year duration, similar to the 'Bologna system' in Europe. This means that the time that students have to complete their studies and compensate for any deficiencies in language, research skills and field knowledge is substantially reduced. This has been particularly highlighted as a problem in the case of China (Cyranoski, 2011), it should be noticed however, that both: Hong Kong and the entire region of China follow this system.

Friedman (2015) makes a lengthy comparison between the 3-year PhD's system—Anglo-Australian/Bologna—and the 5-year system—American. He notices that while the Bologna system has partially achieved the goals of increasing mobility, extending resources, reducing costs and reducing elitism, it all has come at a cost on quality, which are even more aggravated in the case of the Australian and UK universities because these forbid the inclusion of any coursework within the PhD programme. As a result, in the Bologna System students move from practice in BA to 'advanced practice' in Masters, and reach the doctoral stage underprepared in basic skills to support research, like writing (Friedman, 2015).

In contrast, in North America the average PhD lasts from 5 to 6 years—two or three years longer—and it is common that Universities assess incoming students, asking them to go take several courses if deemed necessary. These courses cover field and research methods training, as well as subjects from related disciplines, students also have to pass a number of tests before advancing to candidacy, including language proficiency. The extra 2 years make a substantial difference in the level of skills that students can acquire, and provide more opportunities to make up for the knowledge gaps previously illustrated.

The difference between the systems is even more substantial in institutions following the Bologna system but outside the English-speaking world. While in North American PhDs have more time to make up for language limitations, universities following the Bologna system in countries where the official language is not English are somehow asking students to do advanced writing in a foreign language before they learn to do it in their own, which often results in severe theses problems.

While the PhD program at the school is 3-years long, there is a basic coursework that all PhD students need to complete in order to be eligible for graduation. This coursework includes a Research Methods course, a Design Philosophy course and a Research Ethics course which is mandatory for all research students at PolyU. The school also assesses incoming student's English language proficiency on its own, and in case necessary they are required to attend English training courses. This addresses some of the gaps that incoming students may present, on the other hand however, it shortens the time that students have available to complete their research. One of the things the students asked was how they felt about the coursework; if they would rather keep it or remove it from the study plan, the responses were basically half and half.

To enter the PhD programme at the school is not an admission requirement to hold a previous degree in design, however as the school faculty and staff pointed out, while the requirement as such is not formally written, nearly of the students admitted in the program have either at least one degree, or some career background in the field. And in regards to their English proficiency, the University, as a general requirement, assesses all incoming PhD students for English proficiency on its own. Students found to need help, are then required to attend a series of remedial courses. Another tool that all graduate students in the university have is the 'Effective English for Postgraduate Research Students (EEPRS)' programme, managed by the department of English and aimed at supporting and developing academic writing. The center offers free of charge mentorship and several workshops and activities.

Being the length of the PhD programme 3 years, it is logical to assume that if some students have to take remedial courses, by the time their language skills are ready to write, they would be almost by the end of their normal study period. However, a common way in which students solve this problem is by using their 'associate money' to have their papers proofread and/or copyedited; they can also do the same with their thesis.

This is not to say that there is no room for improvement; the latest 'School of Design Research Retreat Report' highlights that PhD students need additional support to develop their papers, and that research from the school needs to diversify its output to various forms other than text. The report also points out that there seems to be an over-focus on methodology rather than on contribution as well as a lack of clarity in communicating research questions and conclusions. Despite these caveats, the report highlights that according to the latest evaluation of research in Hong Kong, the output of the School of Design, is in general terms of good quality (School of Design, The Hong Kong Polytechnic University, 2015).

Overcoming the Challenges of Building Community

The newness of the field coupled with a knowledge gap prevent a culture of research to flourish across organizations. The fact that experienced researchers are scarce in art & design has an effect on the environment and culture of research, as Friedman (2015) notices in 'Writing for the PhD in Art and Design: Issues for Research Supervisors and Research Students':

Human beings construct and shape cultures through behavior and by encouraging specific forms of behavior. In apprentice programs, apprentices model and adopt the behaviors they observe around them. This creates tacit knowledge. Effective apprentice programs in research require more than a single master. They require an environment populated by skilled journeymen and masters. Schools in the creative and performing arts and design differ from mature research environments because they lack a large population of skilled journeyman researchers and master researchers (Friedman, 2015, p. 22).

Similarly, Coleman & Alford (2010), acknowledge that; "Collaboration cannot take place without working to build professional relationships within the organization." And that the benefits of a collaborative community go beyond mere feedback on research; "Social relationships are necessary so that collaborative reflection is possible" (Coleman & Alford, 2010).

Perhaps an effective way to encourage collaboration and community building amongst the students is through the curriculum. Coleman & Alford (2010), show a good example of how the curriculum can be designed so that students engage in 'collaborative reflection,' in this case, the study plan is designed so that students engage in action research as a group, eventually they have to present their findings as a group as well. This not only strengthens bonds between the group, but provides opportunities for them to provide feedback to each other (Coleman & Alford, 2010).

Building a close community of researchers is perhaps one of the areas that offer the bigger room for improvement. In a previous work, these authors have discussed the opportunities and challenges to facilitate informal learning in China (Siu & Contreras, 2015). The latest research retreat of the school of design report also identified this as an area that needs attention. Amongst other things, the report highlights the need to integrate PhD students into the overall research activities taking place in the school. It also mentions the need to have students organize events on their own and to build-up a stronger community so they can gain more visibility and stop working in isolation (School of Design, The Hong Kong Polytechnic University, 2015). One way in which this is being addressed, is by organizing two PhD Seminars per year. The first 'PhD Seminar Winter Session 2016' took place from the 26 to the 28 of January. In it, PhD students presented their research and got feedback from other students as well as from guest professors. The plan is to hold these seminars twice a year, which should contribute to enhance the issues previously highlighted and hopefully raise enthusiasm towards the 'PhD Wednesdays' initiative described next. In addition, the first student representative was appointed by the end of 2015, with the intention of having a better channel of communication between PhD students and the School Research Committee.

Since September 2015, a group of students started and initiative to hold a monthly to share information and questions about each one's work. It was originally dubbed the 'PhD Wednesdays' because of the plan to hold it in the middle of the week. The idea is to use these gathering as an opportunity to interact with each other and start building a fruitful community and research culture within the programme. The same

Figure 1. One of the PhD students presenting her progress during the 2016 Winter Seminar
(By Authors, 2016)

group recently set up the 'PhD Students of the School of Design Facebook Page,' with the intention of having yet another channel for interaction and a platform to share useful information such as conference and journal calls (PhD Students, School of Design Hong Kong PolyU, 2016).

Overcoming the Challenges of Academia

Academia has often been criticized for working in isolation, (Watts, 2014; Tse & Esposito, 2014). Margolin (2010) as well, points at the disconnection within and between design and other fields as one of the reasons behind the misalignment of PhD programmes out there. Similarly, there is no link between the community of design researchers and those who actually do design; even major professional associations have no relation with design research associations. There is 'no clear relation between the companies that need specialists in the design of objects and complex system, and the universities that should train these specialists' (Margolin, 2010).

This disconnection contributes to exacerbate the 'meta problem' of not knowing what to research. While design research certainly exists, and is being made, unawareness about it persists in the field because much work that could be described as design research is actually carried out by specialists from other fields. Moreover, research done in companies is often confidential, thus the research problems these companies are working on are not known outside their boundaries, and so, the 'nodes' of research activity that should guide research agendas, are still missing and unconnected, limited efforts are done in all directions.

In the case of the school, some labs like the Public Design Lab have a long history of collaborations with the local government, but perhaps other labs interested in industry collaborations may struggle more. One of the reasons is that most manufacturing industries now reside in China, which in a way implies cross-border relations.

The school has also engaged in actions that will aid in building collaboration with other institutions, just recently, the school established a partnership with Swinburne University of Technology in Australia. This partnership will foster community building and collaboration within and outside the school of design. It should also aid to fill any knowledge gaps in both sides while clarifying/enriching a design research agenda. Students enrolled in the joint PhD will benefit from joint supervision, and will receive a degree from both universities (Williams, 2015).

The school is also one of 6 universities from the Chinese mainland and Taiwan that co-organize a 'Design Research PhD Forum,' the latest of which—the third—was held at the School of Design of Hunan University on the 16th and 17th of April 2016. One of the aims of these forums is precisely to improve communication and cooperation amongst PhDs of top universities, as well as improving the quality of academic research and thesis-writing (Hunnan University, 2016).

Figure 2. PhD student Mingjie Zhu presenting her project during the PhD Forum 2016 at Hunan University
(By Authors, 2016)

Overcoming the Challenges of a New Economic and Geopolitical Landscape

For the past 30 years, the Chinese economy has grown at a record pace. During this time graduates of the school of design—many of them Chinese—found themselves next to a double-digit growing economy where job opportunities abounded. This scenario however has most likely changed for the foreseeable future, the last couple of years seem to mark the beginning of new times of more moderate economic growth for China. A sudden decline of jobs in China could substantially affect thousands of PhD graduates from the region, especially when considered against the explosive growth of PhD students in China over the last decade. It is unrealistic to expect that other regions could absorb these graduates as in fact, it has always been more difficult for PhD degree holders from the region to find jobs abroad (Cyranoski, 2011).

Economic Swings and the Job Market

Giving thought to the employability of PhD graduates will have to become an integral part of the future strategy of the PhD programme at the school of design in

Hong Kong. Other countries attempting to build their research and design muscle have struggled to create enough job opportunities amid economic swings and a sluggish economy. Japan for instance, ended up with 18000 unemployed post-doctoral students who were not taken by industry even when the government offered the equivalent to 50 thousand American dollars in subsidy to companies that hired them (Cyranoski, 2011).

A possibility is to make the PhD more 'universally relevant' for a career outside academia, at the moment PhD graduates often lack practical skills needed for a career in industry (McCook, 2011). This is already being addressed by some universities, with programmes that emphasize 'transferable skills' (Cyranoski, 2011). However building up transferable skills requires additional coursework, this would reduce even more the time that students have left for research in a 3-year PhD programme, and have a negative impact on their theses.

Notwithstanding that the programme is still quite young, comparing to other disciplines such as sciences and engineering, and there is not much information available, the job prospects for the school's graduates are in general good. During the first PhD Winter seminar held in January 2016, Professor Ming Xi Tang shared some statistics about how his ex-PhD and Master's students are employed; 14 are in academic jobs, 7 are doing research, and 1 is employed as designer (Tang, 2016).

The idea of making the PhD more relevant for a career outside academia has not been discussed as such at the school, however it was noticed during our interviews, that with the current PhD programme as it is, is possible to pursue a career in either academia or in any environment where design research is made (Tang, Interview, 2015).

FUTURE RESEARCH DIRECTIONS

As the knowledge economy continues to play an increasingly important role, generating knowledge that can be translated into valuable products and services will become equally important. It is in this scenario, that the importance doctoral education must be considered. Given China's current economic transition which places special emphasis on innovation, this is particularly true in the case of design PhDs. But while the general questions that concern PhDs in Design have started to be answered, still little is known about doctoral design education in the Asia pacific region. Consequently, there is need for further research that allows the community to understand its current state, identify problems, address issues, and allow us all to learn from each other.

In addition, a design-research agenda for the region needs to be developed; this should necessarily involve a closer collaboration with industry and other organizations. Only so, design knowledge relevant for the region will be generated, and the employability of PhD students enhanced. This is work that Universities.

CONCLUSION

For 5 decades the School of Design of The Hong Kong Polytechnic University has been the apex of design education locally and a pioneer in the region. In the 21st century, the school continues to take steps towards improving the quality of its research and PhD programme. In this chapter, the emergence, growth and future of the PhD programme at the school of design, has been reviewed against the backdrop of challenges and issues around design research, the future economic landscape of the region, and doctoral education in general. The chapter also discussed the strategies and actions that the school has taken to cope with these challenges in its past, present and future.

As it has been highlighted, Design differs substantially from other fields, this has several repercussions over doctoral design education, some of which are exacerbated in programs following the Bologna system. This chapter has highlighted some of the actions that the school has taken to overcome some of these challenges. It has also been acknowledged, that due the newness of design and design research as fields of study, many questions about doctoral design education remain open; should prospective PhD students for example, be required to hold a degree in Design? These questions will have to be addressed as the discussion to normalize and consolidate doctorates in design continues.

As humans, we bear the responsibility for the material world entirely. Having become 'the factory of the world,' the world's largest population, increasing problems with pollution, and a complicating global economic landscape, China stands at a crucial and complicated time in which Design, as a process concerned with bringing forth the material world, can provide valuable insights to make informed decisions that result in a better materialization of the world around.

It is in this landscape, that, being the cusp of design studies in Hong Kong, a role-model for institutions in the Chinese mainland, and a pioneer of design studies in Asia, the importance of learning about the successes, failures and plans of the PhD programme at the School of Design of The Hong Kong Polytechnic University cannot be overstated. It is the hope of these authors that this small contribution helps to make this possible.

ACKNOWLEDGMENT

We would like to thank the Research Grants Council of Hong Kong for the support through the Hong Kong PhD Fellowship Scheme as well as the support of The Hong Kong Polytechnic University, without which this work would have not been possible. We would also like to thank the students, staff and faculty who kindly participated in the survey and provided us information.

REFERENCES

Abrami, R., Kirby, W., & Warren, M. (2014). *Why China Can't Innovate.* Retrieved from Harvard Business Review: https://hbr.org/2014/03/why-china-cant-innovate

Coleman, C., & Alford, B. (2010). The influence of a doctoral program on growth as a scholar-practitioner leader: Listening to the voices of the graduates. *scholarlypartnerships.edu, 2*(2), 5.

Cyranoski, D. (2011). Education: The PhD factory. *NATNews*, *472*(7343), 276–279. PMID:21512548

De Bont, C. (2015). The Ignite Innovation Program: A Design-Driven Model for Social Innovation from Hong Kong. *International Journal for Cultural and Creative Industries*, *2*(2), 54–67.

Durling, D. (2002). Discourses on research and the PhD in Design. *Quality Assurance in Education, 10*(2), 79–85. doi:10.1108/09684880210423564

Friedman, K. (2015). *Writing for the PhD in Art and Design. Issues for Research Supervisors and Research Students.* Retrieved from Academia.edu: https://www.zotero.org/ncorreia/items/CD6377VA

Hunnan University. (2016). 关于*2016*设计研究博士论坛. Retrieved from School of Design Hunnan University: http://design.hnu.edu.cn/index.php/phdc2016

Law, A. (2011). *Hong Kong Polytechnic University receives major funding for future School of Design.* Retrieved from ico-D: http://www.ico-d.org/connect/index/post/1777.php

Margolin, V. (2010). Doctoral education in Design: Problems and prospects. *Design Issues*, *26*(3), 70–78. doi:10.1162/DESI_a_00031

McCook, A. (2011). 04 20). Rethinking PhDs. *Nature*, *472*(7343), 280–282. doi:10.1038/472280a PMID:21512549

PhD Students, School of Design Hong Kong PolyU. (2016). *PhD Students at The School of Design of the Hong Kong PolyU Facebook Page*. Retrieved from Facebook: https://www.facebook.com/Page-of-the-PhD-Students-at-The-School-of-Design-of-The-Hong-Kong-PolyU-392217560902117/

Rossi, A. (Director). (2014). *Ivory Tower* [Motion Picture].

School of Design, Hong Kong Polytechnic University. (2014). 50th Anniversary Exhibition. Hong Kong: Author.

School of Design, Hong Kong Polytechnic University. (2015). *About PolyU Design.* Retrieved from PolyU Design: http://www.sd.polyu.edu.hk/en/about/about-polyu-design

School of Design, The Hong Kong Polytechnic University. (2015). Hong Kong: School of Design Research Retreat Report.

Siu, K. W., & Contreras, G. (2015). Personal Interaction and Informal Learning: The Case of China. In P. C. O. Mejiuni & P. C. O. Mejiuni (Eds.), *Measuring and Analyzing Informal Learning in the Digital Age* (pp. 140–152). Hershey, PA: IGI Global. doi:10.4018/978-1-4666-8265-8.ch010

Tang, M. X. (2016). Towards an Outcome-Based Postgraduate Education: A perspective for the students. Hong Kong: Academic Press.

Tse, T., & Esposito, M. (2014). *Academia is disconnected from the real world.* Retrieved from Financial Times: http://www.ft.com/cms/s/2/4f5fc7a2-7861-11e3-831c-00144feabdc0.html#axzz3uGO9Zo2H

Watts, A. (2014). *Gruber Thinking in Climate Science: Disconnect Between Academia And The Real World.* Retrieved from wattsupwiththat.com: http://wattsupwiththat.com/2014/12/12/gruber-thinking-in-climate-science-disconnect-between-academia-and-the-real-world/

Williams, A. (2015). *Swinburne and Hong Kong Poly University partner to offer joint PhD program.* Retrieved from Swinburne University of Technology: http://www.swinburne.edu.au/news/latest-news/2015/11/swinburne-and-hong-kong-poly-university-partner-to-offer-joint-phd-program.php

Yeo, J. P.-H. (2012). Research in Art and Design Education: A review. *Academic Research International*, *3*(3), 303.

KEY TERMS AND DEFINITIONS

Bologna System: Also known as 'Bologna Process,' refers to a series agreements between European countries derived from the integration of the European Union, and aimed at reaching standards in higher education qualifications awarded by member states. It was proposed originally at the University of Bologna in 1999, thus its name. It came after the 'Sorbonne declaration' of 1998 in Paris, in which initial member states of the EU committed themselves to "harmonizing the architecture of the European Higher Education system." The process has been opened up to other countries signatory to the European Cultural convention and currently has 50 participating countries.

Cacophony: Literally meaning unpleasant, loud and dissonant sounds, in the context of the chapter the word is used by Victor Margolin to describe the state of 'disharmony' and 'nonconformity' in which PhD programmes in design, as well as the discourses behind them have emerged.

Constructivism: Constructivism is a philosophical understanding regarding the nature of knowing, and suggesting that we (humans) build knowledge from what we experience. For the same reason, it has a direct application to education and learning in general. This philosophical stance, mostly epistemological, was popularized by renown psychologist Jean Piaget in his multiple works about child development.

Chinese Mainland: Although Hong Kong and Macau are part of China, they are considered 'Special Administrative Regions' or 'SAR,' with their own semi-autonomous government and laws. When someone refers to 'The Mainland' in the context of China, it usually means the largest territory, comprising the different provinces, but excluding these SARs and the islands.

Design Research: Originally understood as research into the process of design, the concept has expanded to include research through that process. Aims to understanding and enhance the design processes and practice in general terms, generating insights that can be applied to any particular discipline (i.e., Industrial Design or Graphic Design). It can be understood too as the research that designers do when working on a design project. It has been pointed out that in comparison to other fields, design research is very new and lacks its own methods as well as a clearly defined research agenda.

Knowledge Economy: Refers to the notion that knowledge can be used to generate tangible and intangible products with commercial value. It is in general well accepted that with the rise of 'the information society,' the world economy is in a state of transition to become a 'knowledge economy,' in which case, the importance to education is greater than ever before.

Quacquarelli Symonds (QS): Is a British company that provides services for those looking to study abroad. The company organizes events and produces guides in 35 countries. One of its publications is a global ranking of universities that has been expanding to cover specific subject areas like Design. It has become one of the best known University rankings in the world.

Compilation of References

Abrami, M. R., & Kirby, C. W. (2014, March). Why China Can't Innovate. *Harvard Business Review*. Retrieved from: https://hbr.org/2014/03/why-china-cant-innovate

Abrami, R. M., Kirby, W. C., & McFarlan, F. W. (2014). *Why China Cannot Innovate?* Retrieved from https://hbr.org/2014/03/why-china-cant-innovate

Abrami, R., Kirby, W., & Warren, M. (2014). *Why China Can't Innovate.* Retrieved from Harvard Business Review: https://hbr.org/2014/03/why-china-cant-innovate

Abrami, R., Kirby, W., & Warren, M. (2014). Why China can't innovate. Retrieved from https://hbr.org/2014/03/why-china-cant-innovate

Abrami, R. M., Kirby, W. C., & McFarlan, F. W. (2014). Why China can't innovate. *Harvard Business Review*, *92*(3), 107–111.

AIGA (American Institute of Graphic Arts). (2014). *2014 national median total cash compensation.* Retrieved from http://designsalaries.aiga.org/#salaries-list/

Alonso García, C. (2008). Learning Styles: Present and Future. *Journal of Learning Styles, 1*(1). Retrieved on 25 April, 2016 on: http://learningstyles.uvu.edu/index.php/jls/article/view/73

Amabile, T. M. (1979). Effects of external evaluation on artistic creativity. *Journal of Personality and Social Psychology*, *37*(2), 221–233. doi:10.1037/0022-3514.37.2.221

Amabile, T. M. (1982). Social psychology of creativity: A consensual assessment technique. *Journal of Personality and Social Psychology*, *43*(5), 997–1013. doi:10.1037/0022-3514.43.5.997

Amabile, T. M. (1996). *Creativity in context: update to 'The social psychology of creativity.* Boulder, CO: Westview.

Anonymous. (2014). *About.* Retrieved September 19, 2014, from: http://en.nua.edu.cn/s/92/t/585/p/21/c/1863/list.htm

Anonymous. (2016). *Organizing Your Social Sciences Research Paper: Theoretical Framework.* Retrieved May 11, 2016, from: http://libguides.usc.edu/writingguide/theoreticalframework

Anonymous. (Essay), translation by Allen-Ebrahimian, B. (2015). *Why do Chinese lack creativity?* Retrieved from http://foreignpolicy.com/ 2015/06/23/ china-innovation-creativity-research-patents/

Art News. (2012). What will bring after upgrading Arts Category? The meeting of history and current situation of the development of art discipline. *Arts News*. Retrieved April 16, 2016 from http://www.zgnfys.com/a/nfms-1978.shtml

Atalay, M., Anafarta, N., & Sarvan, F. (2013, April). The Relationship between Innovation and Firm Performance: An Empirical Evidence from Turkish Automotive Supplier Industry. *Procedia: Social and Behavioral Sciences*, *75*(3), 226–235. doi:10.1016/j.sbspro.2013.04.026

Atkinson, D. (1997). A Critical Approach to Critical Thinking in TESOL. *TESOL Quarterly*, *31*(1), 71–94. doi:10.2307/3587975

Atkinson, D. (1999). TESOL and Culture. *TESOL Quarterly*, *33*(4), 625–654. doi:10.2307/3587880

Avril, E. (2009). *The art of China's cultural revolution.* Herbert F. Johnson Museum of Art, Cornell University. Retrieved from: http://museum.cornell.edu/exhibitions/art-chinas-cultural-revolution

Bachelard, B. (1942). *Water and Dreams: An Essay on Imagination of Matter* (E. Salsali, Trans.). Paris, France: Corti.

Baer, J. (1994). Performance assessments of creativity: Do they have long-term stability? *Roeper Review*, *17*(1), 7–11. doi:10.1080/02783199409553609

Bai, M. (2010). *The Future of "Made in China"*. Retrieved from http://www.chinatoday.com.cn/ ctenglish/se/txt/2011-12/30/content_417573.htm China Today.

Baker, A. (2012). Integral sustainable design: Transformative perspectives, by Mark Dekay: Book Review. *Journal of Ecological Design*. Retrieved March 15, 2016 from http://www.ecotecture.com/integral-sustainable-design-transformative-perspectives-by-mark-dekay/

Barboza, D. (2006). *China Says One of Its Scientists Faked Computer Chip Research*. Retrieved from http://www.nytimes.com/2006/05/14/technology/ 14chip.html?_r=2&

Barnett, C. (2013). *Top 10 Crowdfunding Sites For Fundraising*. Retrieved November 11, 2014, from: http://www.forbes.com/sites/chancebarnett/2013/05/08/top-10-crowdfunding-sites-for-fundraising/

Baudrillard, J. (1994). *Simulacra and Simulation.* University of Michigan Press (first published 1981 in French).

Bauman, Z. (2008). *The Art of Life*. Polity Editions.

Beghetto, R. A. (2007). Does creativity have a place in classroom discussions? Prospective teachers' response preferences. *Thinking Skills and Creativity*, *2*(1), 1–9. doi:10.1016/j.tsc.2006.09.002

Bell, J. (1999). *Doing your research project: a guide for first researchers in education and social science*. Academic Press.

Bell, K. (2014). 7 things copycat Xiaomi blatantly ripped off from Apple. Retrieved from http://www.cultofandroid.com/66569/xiaomi-dont-copy-apple/

Bermejo Velasco, P. E., & Izquierdo Megía, R. J. (2011). *Cerebro rico, cerebro pobre (Brainrich, Brainpoor)*. Cultiva Libros.

Biggs, J. (1996). Western misperceptions of the Confucian-heritage learning culture. In D. Watkins & J. Biggs (Eds.), *The Chinese learner: Cultural, psychological, and contextual influences* (pp. 45–67). Hong Kong: CERC and ACER.

Biggs, J. (1998). Learning from the Confucian heritage: So size doesn't matter? *International Journal of Educational Research*, *29*(8), 723–738. doi:10.1016/S0883-0355(98)00060-3

Biggs, J. B. (1989). Approaches to the enhancement of tertiary teaching. *Higher Education Research & Development*, *8*(1), 7–25. doi:10.1080/0729436890080102

Biggs, J., & Watkins, D. (1996). *The Chinese learner in retrospect*. Melbourne: CERC, Hong Kong, and ACER.

Bjögvinsson, E., Ehn, P., & Hillgren, P. (2012). Design things and design thinking: Contemporary participatory design challenges. *Design Issues*, *28*(3), 101–116. doi:10.1162/DESI_a_00165

Bloedon, R. V., & Stoke, D. R. (1994). Making University-industry Collaborative Research Succeed. *Research Technology Management*, *37*, 44–49.

Bloom, B. S., & David, R. K. (1956). *Taxonomy of educational objectives: The classification of educational goals, by a committee of college and university examiners*. New York: David McKay Publications.

Bloomberg, L. D., & Volpe, M. (2012). *Completing your qualitative dissertation: A road map from beginning to end*. Sage Publications.

Boden, M. A. (2004). *The creative mind: Myths and mechanisms*. Psychology Press.

BoF. L.JingY. Y.LiuW. (薄芙丽, 景怀宇, & 刘玮) (2009). 探索实践教学新路 优化人才培养规格——关于中国动画教育改革的设想. 装饰. 6, 042.

Bol, P. (1992). *'This culture of ours'. Intellectual transitions in T'ang and Sung China*. Stanford, Ca.: Stanford University Press.

Bottge, B., & Osterman, L. (1998). Bring the Workplace to Classroom. *Educational Leadership*, *55*, 76–77.

Brown, T. (2008). *Tales of creativity and play.* Presentation at the Serious Play Conference. Retrieved October 31, 2015, from http://www.ted.com/talks/tim_brown_on_creativity_and_play.html

Brown, T. (2008). Design Thinking. *Harvard Business Review*, 84–94. PMID:18605031

Brown, T. (2008). Design Thinking. *Harvard Business Review*, 84–96. PMID:18605031

Bruneel, J., D'Este, P., & Salter, A. (2010). Investigating the factors that diminish the barriers to university-industry collaboration. *Research Policy*, *39*(7), 858–868. doi:10.1016/j.respol.2010.03.006

Buci-Glucksman, C. (2001). *The Esthetic of Time in Japan: From Zen to the Virtual* (E. Salsali, Trans.). Paris, France: Galilée.

Bullis, K. (2013). *How Tesla Is Driving Electric Car Innovation.* Retrieved from https://www.technologyreview.com/s/516961/how-tesla-is-driving-electric-car-innovation/

Caillois, R. (2001). *Man, play and games.* Urbana, IL: University of Illinois Press.

Caine, R., & Caine, G. (1991). *Making Connections:Teaching and the human brain.* Alexandria, VA: ASCD.

Cain, S. (2012). *The Power of Introverts in a World that Can't Stop Talking*. New York, NY: Crown.

Caldera, A., & Debande, O. (2010). Performance of Spanish universities in technology transfer: An empirical analysis. *Research Policy*, *39*(9), 1160–1173. doi:10.1016/j.respol.2010.05.016

Caldwell-Harris, C., & Aycicegui-Dinn, A. (2008). *Emotion and lying in a non-native language*. Elsevier.

Candy, L., & Edmounds, E. (2005). *Interacting: Art, Research and Creative Practitioner*. Faringdon, United Kingdom: Libri.

Cao, L. (2005). *Chinese decorative arts Traditional and its contemporary culture value*. (PhD thesis). Chinese National Academy of Arts. Retrieved February 20, 2016 from http://cdmd.cnki.com.cn/Article/CDMD-84201-2006060477.htm

Chang, G. (2015). The Chinese 'century' is already over. Retrieved from The National Interest website: http://nationalinterest.org/feature/the-chinese-century-already-over-12915

Chang, K. M. (2000). How to Promoting University-industry Cooperation to Enhance Design Capability. *Symposium Proceeding on Design Education of Japan and Asia*. YunTech.

Chen, B. (2014, October 10). Apple's Jony Ive has harsh words for Xiaomi. *The New York Times*. Retrieved from http://bits.blogs.nytimes.com/2014/10/10/apples-jony-ive-has-harsh-words-for-xiaomi/

Chen, H. & Ma, L. T. (2011). Reflection and prospect of the higher education administration reform. *Journal of National Academy of Education Administration*, (6).

Chen, J. L. Zhang, F. X (2011). A Study of S&T Investment Strength and Development Tendency in Foreign Countries. *International Journal of Business Administration, 2*(2).

Chen, K. S., & Lin, T. Y. (1990). OEM? Or own brands? *Taiwan Economic Research Monthly*, 134.

Chen, X. H. (2007). Between craft and design: The modernity course of the 20 century Chinese art design. Chongqing University Publishing House.

Chen, Y. C. (2007). *A Study on Business Strategy of Small and Medium Enterprise in Taiwan for the Transformation of OEM/ODM to OBM*. (MS Thesis). Department of Industrial Management National Taiwan University of Science and Technology.

Chen, Z. H. (2000). The Review and Improvement of Industry-academia Cooperation in Research Program. *Science Development*, 921-927.

Chen, H. C. (2001). The Practical Method of Industry-academic Cooperation. *Journal of Technological and Vocational Education.*, *61*, 57–61.

Chen, T. C., Lin, B. W., & Chu, M. A. (2012). Knowledge Exploration and Exploitation in the University-Industry Collaboration. *NTU Management Review*, *22*(2), 59–86.

Chick, A. (2012). Design For Social Innocation: Emerging Principles And Approaches. *Iridescent-Icograda Journal of Design Research, 2*(1), 56-60.

Child, I. L. (1972). Esthetics. *Annual Review of Psychology*, *23*(1), 669–694. doi:10.1146/annurev.ps.23.020172.003321

China Education Daily. (2016). The first higher education quality national report. *China Education Daily*. Retrieved April 18, 2016 from http://www.moe.edu.cn/jyb_xwfb/s5147/201604/t20160408_237162.html

China Education Network Channel. (2011). Arts upgraded to an independent discipline to become the thirteenth disciplines. *China Education Network Channel*. Retrieved December 15, 2015 from http://www.designedu.cn/showposts.aspx?id=299&columnid=7&channelid=1

Chinese Design Education. (2015). Chinese design colleges and universities and courses. *China Design Education*. Retrieved December 15, 2015, from http://www.designedu.cn/curriculumsearch.aspx?channelid=2&columnid=24&selProf=0&selarea=all&seldeg=0&stext=&page=21

Chiu, C., & Hong, Y. (2006). *Social psychology of culture*. New York: Psychology Press.

Chou, T. D. (2013). Taiwanese Strategic Mode to Promote Industry-academia Partnership—Industry-academia R&D and Talent Cultivation. *The Journal of Chaoyang University of Technology*, *18*, 85–109.

Christensen, C. (2011). *The Innovator's Dilemma: The Revolutionary Book That Will Change the Way You Do Business*. Harper Business. doi:10.15358/9783800642816

Chung, C. J. (2015). *An Analysis of Key Factors for University Faculty to Participate in Industry-academic Cooperation– Applying Analytical Hierarchy Process Method.* (MS Thesis). Graduate School of Visual Communication Design, Kun Shan University.

Chu, T. (2006). *Design based on insect: Experimental teaching & learning of city design school, Central Academy of Fine Arts*. Beijing: China Textile Publishing House.

Clark, H. (2009). Back to the future, or forward? Hong Kong design, image, and branding. *Design Issues*, *25*(3), 11–29. doi:10.1162/desi.2009.25.3.11

Cohen, R. (2014). *Design Thinking: A Unified Framework For Innovation*. Retrieved from http://www.forbes.com/sites/reuvencohen/2014/03/31/design-thinking-a-unified-framework-for-innovation/#6fa6dcf856fc

Coleman, C., & Alford, B. (2010). The influence of a doctoral program on growth as a scholar-practitioner leader: Listening to the voices of the graduates. *scholarlypartnerships.edu,* *2*(2), 5.

College Art Association. (2015). MFA Standards: Statement on Terminal Degrees in the Visual Arts and Design. *College Art Association.* Retrieved February 18, 2016, from http://www.collegeart.org/guidelines/mfa

Collin, R. (1998). *The Sociology of Philosophies: a Global Theory of Intellectual Change*. Harvard University Press.

Contreras, G., & Siu, K. (2016). *Design Education in China: Fostering creating and innovation in China and emerging nations*. Retrieved from: http://gicoga.wix.com/design-edu-in-china#!introduction/cee5

Crespi, G., D'Este, P., Fontana, R., & Geuna, A. (2011). The impact of academic patenting on university research and its transfer. *Research Policy*, *40*(1), 55–68. doi:10.1016/j.respol.2010.09.010

Cropley, A. J. (1992). *More ways than one: Fostering creativity*. Ablex Publishing.

Cropley, A. J. (2001). *Creativity in education & learning: A guide for teachers and educators*. Psychology Press.

Cross, N. (2001). Designerly ways of knowing: Design discipline versus design science. *Design Issues*, *17*(3), 49–55. doi:10.1162/074793601750357196

Cross, N. G. (1982). Designerly ways of knowing. *Design Studies*, *3*(4), 221–227. doi:10.1016/0142-694X(82)90040-0

Cross, N. G. (1990). The nature and nurture of design ability. *Design Studies*, *11*(3), 127–14. doi:10.1016/0142-694X(90)90002-T

CsikszentInihalyi, M. (1999). Implications of a systems perspective for the study of creativity. In R. J. Sternberg (Ed.), Handbook of creativity (pp. 313-335). New York: Cambridge University Press.

Cyranoski, D. (2011). Education: The PhD factory. *NATNews*, *472*(7343), 276–279. PMID:21512548

D. K. (2014, September 14). *Q. and A.: Yong Zhao on Education and Authoritarianism in China.* Retrieved from The New York Times blog.

D'Angelo, E. (1971). *The Teaching of Critical Thinking*. Amsterdam: B. R. Grunner.

Dahl, J. (2016). Chinese government luring 'sea turtles' home to launch startups. Retrieved from Forbes website: http://www.forbes.com/sites/jordyndahl/2016/05/03/chinese-government-luring-sea-turtles-home-to-launch-startups/#636e061b478a

Dai, X. X. (2000). *The Popularization and Marketization of Higher Education*. Taiwan.

Damasio, A. (2005). *Descartes Error: Emotion, Reason and the Human Brain*. Penguin Books.

Danxi, S. (2015, April 21). *China's new Gaokao reform and its global implication*. Retrieved from Shorelight Education Blog: http://blog.shorelight.com/chinas-new-gaokao-reform-and-its-implication-on-international-admissions

Darwin, C. (1859) The origin of species. Wordsworth.

Davidson, B. (1998). Critical thinking faces the challenge of Japan. *Inquiry*, *14*(3), 41–53.

Dawson, V. I. (1997). In search of the wild bohemian: Challenges in the identification of the creatively gifted. *Roeper Review*, *19*(3), 148–152. doi:10.1080/02783199709553811

Day, R. (2003). Teaching critical thinking and discussion. *Language Teaching*, *27*(7), 25–27.

De Bont, C. (2015). The Ignite Innovation Program: A Design-Driven Model for Social Innovation from Hong Kong. *International Journal for Cultural and Creative Industries*, *2*(2), 54–67.

De Saussure, F. (2013). *Course in general linguistics*. London: Bloomsbury Publishing PLC.

Debord, G. (1970). *The society of the spectacle*. Detroit, MI: Black & Red.

DeKay, M. (2011). *Integral sustainable design: transformative perspectives*. Routledge.

Department for Culture. Media and Sport. (1998). Creative industries mapping document. Author.

Dewey, J. (1927). *The Public and its Problems*. Athens, OH: Ohio University Press.

Doin, G. (Director). (2012). *The Forbidden Education* [Motion Picture].

Don, C. (2014). *From 'Made in China' to 'Designed in China'*. Paper presented at the M+ Matters, Hong Kong. Retrieved from http://www.mplusmatters.hk/importexport/pdf/herenke.pdf

Dorst, K., & Reymen, I. (2004). Levels of Expertise in Design Education.*2nd International Engineering and Product Design Education Conference (IEPDE).*

Drake, D. (2014). *Crowdfunding: It's No Longer a Buzzword.* Retrieved November 11, 2014, from: http://www.crowdsourcing.org/editorial/crowdfunding-its-no-longer-a-buzzword/32268

Dunne, A. (1999). Hertzian tales: electronic products, aesthetic experience and critical design. London, UK: Royal College of Art Computer-Related Design Research Studio.

Durkheim, E. (1982). The rules of sociological method. The Free Press. doi:10.1007/978-1-349-16939-9

Durling, D. (2002). Discourses on research and the PhD in Design. *Quality Assurance in Education, 10*(2), 79–85. doi:10.1108/09684880210423564

Economist, T. (2015). The economy: Go slow. Retrieved from The Economist website: http://www.economist.com/news/china/21645839-prime-minister-seeks-lower-expectations-economy-go-slow

Economist. (2007). *Lessons from Apple: What other companies can learn from California's master of innovation.* Retrieved from http://www.economist.com/node/9302662

Edwards, J. (2015). China's massive economic advantage over the world is about to disappear. Retrieved from Business Insider website: http://uk.businessinsider.com/chinas-four-decade-miracle-ends-this-year-2015-8

Edwards, M., McGlodrick, C., & Oliver, M. (2006). Creativity and curricula in higher education: Academics' perspective. In N. Jackson, M. Oliver, M. Shaw, & J. Wisdom (Eds.), Developing creativity in higher education: An imaginative curriculum. Routledge.

Eisenhardt, K. M. (1989). Building theories from case study research. *Academy of Management Review, 14*(4), 532–550.

Eker, T. H. (2007). *Secrets of the Millionaire Mind: Think rich to get rich.* Piatkus.

Elder, L. (n.d.). *Using Critical Thinking to Get Beyond Groupthink.* Foundation for Critical Thinking. Retrieved from: http://www.criticalthinking.org/files/The%20Emancipated%20Mind%20FINAL.docx

Elder, L., & Paul, R. (1996). Critical thinking: A stage theory of critical thinking. *Journal of Developmental Education, 20*(1), 34–35.

Elliot, M. (2008, January 17). A tale of three cities. *Time Magazine.*

Ennis, R., & Weir, E. (1985). *The Ennis-Weir Critical Thinking Essay Test*. Pacific Grove, CA: Critical Thinking Press and Software.

Eno, B., & Schmidt, P. (1975). *Oblique strategies: over one hundred worthwhile dilemmas*. Retrieved November 2, 2015, from http://www.rtqe.net/ObliqueStrategies/

Esbjorn-Hargens, S. (2009). An all-inclusive framework for the 21st century: an overview of integral theory. *Integralpost*. Retrieved December 10, 2015 from https://integrallife.com/integral-post/overview-integral-theory

Esbjorn-Hargens, S. (2010). Introduction. In Integral theory in action: Applied, theoretical, and constructive perspectives on the AQAL model. Albany, NY: State University of New York Press.

Feldman, M. P., & Florida, R. (1994). The geographic sources of innovation: Technological in frastructure and product innovation in the United States. *Annals of the Association of American Geographers*, *84*(2), 210–229. doi:10.1111/j.1467-8306.1994.tb01735.x

Feng, W. W., & Siu, K. W. M. (2009). Meeting the challenges of education reform: Curriculum development of technology education in mainland China and Hong Kong. In *International Conference on Technology Education in the Asia Pacific Region Conference Proceedings: Less is More* (pp. 447-457). Taipei: National Taiwan Normal University.

Ferrero, G. P. (2002). *The Cultural dimension of international business*. Pearson Educational Inc.

Fleming, B. (2013). *Design education for a sustainable future*. Routledge/Earthscan.

Fletcher, A. (2001). *The art of looking sideways*. London, UK: Phaidon.

Freedman, E. B. (2007). Is teaching for social justice undemoctatic? *Harvard Educational Review*, 77.

Friedman, K. (2015). *Writing for the PhD in Art and Design. Issues for Research Supervisors and Research Students.* Retrieved from Academia.edu: https://www.zotero.org/ncorreia/items/CD6377VA

Fry, T. (2003). The 'futurings' of Hong Kong. *Design Issues*, *19*(3), 71–82. doi:10.1162/074793603768290856

Fung, A. S. Y., & Lo, A. Y. N. (2001). Design education in China: New proposals to address endemic problems. *Journal of Art & Design Education*, *20*(2), 171–179. doi:10.1111/1468-5949.00264

Fung, A., Lo, A., & Rao, M. N. (2005). *Creative tools*. Hong Kong: The Hong Kong Polytechnic University.

Gao, F. (2001). The development of China's modern art and design in the 2nd half of 20 century. *Arts Observation*, (6), 54.

Gardner, H. (1993). Multiple Intelligences: The theory in practice. Basic Books.

Gaver, W. W., Bowers, J., & Boucher, A. (2004). The drift table: designing for ludic engagement. *Proceedings of the CHI EA '04 Conference*. ACM Press. doi:10.1145/985921.985947

Geertz, C. (1973). *The Interpretation of Cultures: Selected Essays*. Basic Books.

Ghiselli, E. E., & Siegel, J. P. (1972). Leadership and managerial success in tall and flat organization structures. *Personnel Psychology*, *25*(4), 617–624. doi:10.1111/j.1744-6570.1972.tb02304.x

Goldberg, E. (2001). *The Executive Brain*. Oxford University Press.

Goleman, D. (2005). *Emotional Intelligence: Why it can matter more than IQ*. Bantam Books.

Greeberg, M. D., Hui, J., & Gerber, E. (2013). Crowdfunding: A resource exchange perspective. In *Proceedings of CHI '13 Extended Abstracts on Human Factors in Computing Systems*.

Grignon, P. (2012). *Money as Debt*. Fábula de Goldsmith. Retrieved from http://. asisonlascosas/el-dinero-es-deuda.

Guastello, S. J. (2008). Creativity and personality. In T. Richards, M. A. Runco, & S. Moger (Eds.), *The Routledge companion to creativity* (pp. 267–278). Routledge. doi:10.4324/9780203888841.ch22

Guilford, J. P. (1950). Creativity. *The American Psychologist*, *5*(9), 444–454. doi:10.1037/h0063487 PMID:14771441

Guilford, J. P. (1970). Creativity: Retrospect and prospect. *The Journal of Creative Behavior*, *5*, 77–87. doi:10.1002/j.2162-6057.1971.tb00877.x

Guo, S., & Blanchard, J. F. (2008). *Harmonious word – and China's new foreign policy*. Lexington Books.

Guy, T. C. (1999). Culture as context for adult education: The need for culturally relevant adult education. In New directions for adult and continuing education: Providing culturally relevant adult education (p. 7). San Francisco: Jossey-Bass. doi:10.1002/ace.8201

Hai, J., & Siu, K. W. M. (2011). Design for all: Social innovation and service design education and practices in China. In P. L. P. Rau (Ed.), *Internationalization, design and global development* (pp. 514–523). Heidelberg: Springer. doi:10.1007/978-3-642-21660-2_58

Han, Y. Y. (2010). Innovation of contemporary art design and inheritance of Chinese traditional culture. *Journal of Changchun University of Technology,* (5).

Handy, C. (2004). *The new alchemists: How visionary people make something out of nothing*. London: Hutchison.

Hang, J. & Cao, X.O. (2009). Design: another enlightenment - evolution of design ideas and practice of 30 years reform and opening up. *Literature & Art Studies*, (1).

Hennessey, B. A., & Amabile, T. M. (1988). The conditions of creativity. In R. Sternberg (Ed.), *The nature of creativity* (pp. 11–38). New York, NY: Cambridge University Press.

Herrmann, N. (1989). The Creative Mind. Brain Books.

Heskett, J. (2014). *Are we entering an era of neuromanagement?* Retrieved from http://hbswk.hbs.edu/item/are-we-entering-an-era-of-neuromanagement

Heskett, J. (2005). *Design: a very short introduction.* Oxford University Press. doi:10.1093/actrade/9780192854469.001.0001

Highet, G. (1950). The art of teaching. Alfred A. Knopf, 194. In P. Rand, (1965) *Design and the play instinct.* In G. Kepes (Ed.), *Education of vision* (p. 156). New York, NY: George Braziller Inc.

HKSARG. (2011). Thematic report: ethnic minorities. In *The 2011 population census.* Census and Statistics Department, The Hong Kong Special Administrative Region Government. Retrieved November 8, 2015, from http://www.census2011.gov.hk/pdf/EM.pdf#Page=28

HKTDC. (2016). Glossary: "Made in China 2025". Retrieved from Hong Kong Trade Development Council website: http://economists-pick-research.hktdc.com/business-news/article/Glossary/Glossary-Made-in-China-2025/glossary/en/1/1X000000/1X0A5GQY.htm

Ho, D. Y. F., & Ho, R. T. H. (2008). Knowledge is a Dangerous Thing: Authority Relations, Ideological Conservatism, and Creativity in Confucian-Heritage Cultures. *Journal for the Theory of Social Behaviour, 38*(1).

Ho, D., & Ho, R. (2008). Knowledge is a dangerous thing: Authority relations, ideological conservatism, and creativity in Confucian heritage cultures. *Journal for the Theory of Social Behaviour, 38*(1), 67–86. doi:10.1111/j.1468-5914.2008.00357.x

Hodgson, G. (2001). Socio-economic consequences of the advance of complexity and knowledge. *Voprosy Economiki, 8.*

Hofstede, G. (2006). Dimensions of Culture. *Journal of Cross-Cultural Psychology, 15*(4), 417-433.

Hong Kong Arts Development Council. (2000). Introduction to Creative Industries: The Case of United Kingdom and Implementation Strategies in Hong Kong. Hong Kong: Author.

Hsiao, S. W., & Chou, J. R. (2004). A creativity-based design process for innovative product design. *International Journal of Industrial Ergonomics, 34*(5), 421–443. doi:10.1016/j.ergon.2004.05.005

Hu, G. H. (1996). Research of Chen Zhifuo craft thought. *Journal of Nanjing Academy of Arts,* (3), 75.

Hudson, A. (2014). *Howard Gardner's Learning Styles.* Retrieved from: http://www.ehowenespanol.com/estilos-aprendizaje-howard-gardner-sobre_327898/

Hui, J., Greenberg, M., & Gerber, E. (2013). Understanding crowdfunding work: implications for support tools. In *Proceedings of CHI '13 Extended Abstracts on Human Factors in Computing Systems.*

Hui, L. (2015, June 9). Made in China 2025: How Beijing is revamping its manufacturing sector. *South China Morning Post.* Retrieved from http://www.scmp.com/tech/innovation/article/1818381/made-china-2025-how-beijing-revamping-its-manufacturing-sector

Huibert, D. M. (2005). *Teaching Management to Chinese Students: a few notes from theory and practice.* Amsterdam: University of the Netherlands.

Huizinga, J. (1972). *Homo Ludens: A study of the play element in human culture.* Boston, MA: Beacon Press.

Hunan University. (2005). *School of Design, Hunan University, School introduction.* Retrieved April 20, 2016 from http://design.hnu.edu.cn/index.php/about/introduction

Hunnan University. (2016). *关于2016设计研究博士论坛*. Retrieved from School of Design Hunnan University: http://design.hnu.edu.cn/index.php/phdc2016

IDEO. (2003). *IDEO method cards: 51 ways to inspire design*. William Stout.

Industrial Design Centre. (2009). *Design as a strategy for developing economy*. Bombay, India: Indian Institute of Technology.

International Council of Design. (2009). *Central Academy of Fine Arts recognized by BusinessWeek*. Retrieved from http://www.ico-d.org/connect/index/post/1365.php

International Organization for Standardization. (n.d.). *China (SAC)*. Retrieved from: http://www.iso.org/iso/about/iso_members/iso_member_body.htm?member_id=1635

Jackson, N., Oliver, M., Shaw, M., & Wisdom, J. (Eds.). (2006). *Developing creativity in higher education: An imaginative curriculum*. Routledge.

Jaswal, V. K., & Neely, L. A. (2006). Adults don't always know best: Preschoolers use past reliability over age when learning new words. *Psychological Science*, *17*(9), 757–758. doi:10.1111/j.1467-9280.2006.01778.x PMID:16984291

Jiang, L. (2003). *The Changes of the chair: Overpassing the concepts*. Beijing: People's Fine Arts Publishing House.

Jiangnan University. (n.d.). *School of Design, Jiangnan University, Baike Baidu*. Retrieved April 20, 2016 from http://www.jiangnan.edu.cn/

Jin, Y. (2007). *Chinese design art theory development research after 1980s*. (PhD thesis). Wuhan University of Technology.

Jones, P., Comfort, D., Eastwood, I., & Hillier, D. (2004). Creative industries: Economic contributions, management challenges and support initiatives. *Management Research News*, *27*(11/12), 134–145. doi:10.1108/01409170410784644

Jubert, M. M. (2001). The art of creative teaching: NACCCE and beyond. In A. Craft, B. Jeffrey, & M. Leibling (Eds.), *Creativity in education* (pp. 17–34). London, UK: Continuum.

Justice, L. (2006). Product design education in China. Retrieved from IDSA website: http://www.idsa.org/sites/default/files/nec06_justice_lorraine.pdf

Kahneman, D. (2011). *Thinking fast and slow*. N.Y. Farrar, Straus & Ciroux.

Kahneman, D., Tversky, A., & Slovic, P. (1982). *Judgement under uncertainty: Heuristic and biases*. Cambridge University Press. doi:10.1017/CBO9780511809477

Kalama Sutta, The Buddha's Charter of Free Inquiry. (2013). Retrieved by: http://www.chinabuddhismencyclopedia.com/en/index.php/Kalama_Sutta

Kaufman, J. C., & Sternberg, R. J. (Eds.). (2006). *The international handbook of creativity*. New York: Cambridge University Press. doi:10.1017/CBO9780511818240

Keane, M. (2013). *Creative industries in China: Art, design and media*. John Wiley & Sons.

Keeley, L., Walters, H., Pikkel, R., & Quinn, B. (2013). *Ten types of innovation: the discipline of building breakthroughs*. New York, NY: Wiley.

Kennedy, S. (2015). Made in China 2025. Retrieved from Center for Strategic and International Studies website: http://csis.org/publication/made-china-2025

Kim, K. H. (2007). Exploring the interactions between Asian culture (Confucianism) and creativity. *The Journal of Creative Behavior*, *41*(1), 28–54. doi:10.1002/j.2162-6057.2007.tb01280.x

Kim, K. H., Lee, H. E., Chae, K. B., Anderson, L., & Laurence, C. (2011). Creativity and Confucianism among American and Korean educators. *Creativity Research Journal*, *23*(4), 357–371. doi:10.1080/10400419.2011.621853

Kolko, J. (2008). The Tenuous Relationship Between Design and Innovation. *Routledge's Artifact, 1*(3), 198-203.

Krishnan, V., & Ulrich, K. T. (2001). Product development decisions: A review of the literature. *Management Science*, *47*(1), 1–21. doi:10.1287/mnsc.47.1.1.10668

Kristiansen, P., & Rasmussen, R. (2014). *Building a better business using the Lego serious play method*. New York, NY: Wiley.

Kuppuswamy, V., & Bayus, B. L. (2014). *Crowdfunding creative ideas: The dynamics of project backers in Kickstarter.* UNC Kenan-Flagler Research Paper, (2013-15).

Lane, C. (2012). *Multiple Intelligences*. Retrieved on 29 November, 2015 from: http://www.tecweb.org/styles/ gardner.html

Law, A. (2011). *Hong Kong Polytechnic University receives major funding for future School of Design.* Retrieved from ico-D: http://www.ico-d.org/connect/index/post/1777.php

Leclerc, R. (2005). 22°15'N, 114°10'E. *Urban China, 5*. (first published 1999 in Tofu Mag-Zine)

Leclerc, R., & Wan, B. (2012). Design play-interactive education toolkit for design project planning and management. In *Design Ed Asia 12 Conference Proceedings.* Hong Kong: School of Design, The Hong Kong Polytechnic University.

Lee, T. (2000). *Education in traditional China. A history.* Leiden: E.J. Brill.

Lewis, R. (2009). *Understanding Pupil Behaviour: Classroom management techniques for teachers.* London: Routledge.

Li, H. (李海玲) (2010). Report on the Higher Animation Education in China (中国动画高等教育发展报告). In China Animation Yearbook 2010 (pp. 469-479). Communication University of China Press.

Li, J. (2010). *Talents-training mode research of higher art education.* (PhD thesis). Hunan Normal University. Retrieved February 10, 2016 from http://cdmd.cnki.com.cn/Article/CDMD-10542-2010127656.htm

Li, M. (2015). *35 years of China modern products design since 1978.* (Master thesis). China Art Research Institute. Retrieved February 8, 2016 From http://cdmd.cnki.com.cn/Article/CDMD-84201-1015397830.htm

Li, W. H., Rosanna, S., & Angelina, S. (2007). *Creative Industries: Singapore and Hong Kong – A Review of Design Initiatives with Implications for the Nurturing of Design Talent.* Retrieved from http://www.ln.edu.hk/mcsln/20th_issue/feature_01.shtml

Li, Y. Z. (2002). Outline of arts and crafts. Shandong Education Publishing House.

Li, Y. Z. (2003). Establishing new type of art design education system, developing the course of Chinese art design education. *2001 Tsinghua International Industry Design Forum - Paper Collection of the National Industry Design Education Seminar.* Beijing: Tsinghua University Publishing House.

Littlewood, W. (2000). Do Asian students really want to listen and obey? *ELT Journal, 54*(1), 31–36. doi:10.1093/elt/54.1.31

Liu, G. Z. (1988). History how to inform future. *Zhuangshi,* (1).

Liu, G. Z. (1987). *New form of counterparty culture – industry design*. Beijing: Exhibition Design Society.

Liu, S. X. (2016). Innovation design: Made in China 2025. *Design Management Review*, *27*(1), 52–58.

Longman, D. G., Atkinson, R. H., & Breeden, J. A. (1997). *Strategic thinking and reading*. Boston: Heinle & Heinle Publishers.

Lopez, L. (2015). This dazzling visualization shows how everything in China is shrinking. Retrieved from Business Insider website: http://www.businessinsider.com/visualization-china-shrinking-economy-2015-8

Lopez, L. (2016). Old China is drowning. Retrieved from http://finance.yahoo.com/news/old-china-drowning-173858576.html

Lorraine, J. (2012). *China's Design Revolution*. Cambridge, MA: The MIT Press.

Louie, K. (2005). Gathering cultural knowledge. Useful or use with care? In J. Rand & J. Carroll (Eds.), *Teaching international students. Improving learning for all* (pp. 17–25). Abingdon, UK: Routledge.

Lu, P. T. (1996). Design and decoration: necessary tension – exploring educational road of Central Academy of Arts & Design. *Zhuangshi,* (5).

Lu, R. (2015). China, start-up nation. Retrieved from http://foreignpolicy.com/2015/04/22/china-startup-nation-from-factory-to-tech-incubator/

Ma, Y. (2008). Visual innovation of contemporary art design based on Chinese traditional culture. *Art Exploration*, (4).

Mac Lean, P. (1990). *The Triune Brain in Evolution: Role in Paleocerebral functions*. Springer.

Magnier, M. (2016, Januray 19). China's economic growth in 2015 is slowest in 25 Years. *The Wall Street Journal*. Retrieved from http://www.wsj.com/articles/china-economic-growth-slows-to-6-9-on-year-in-2015-1453169398

Mahbubani, K. (2008). *The new Asian hemisphere: the irresistible shift of global power to the East*. PublicAffairs.

Mantzanakis, S. C. Epaminondas. (2015). *China 2025 research and innovation landscape*. Retrieved from??????? Rapoza, K. (2015). World Bank States The Obvious On China, China Gets Mad. Retrieved from www.forbes.com/sites/kenrapoza/2015/07/19/world-bank-states-the-obvious-on-china-china-gets-mad/#16406d1f1097

Manu, A. (2007). *The imagination challenge: points of departure for strategic creativity and innovation*. New Riders. Retrieved November1, 2015, from http://english.360elib.com/datu/H/EM315566.pdf

Manu, A. (1998). *Tool toys: tools with an element of play*. Copenhagen, Denmark: Danish Design Centre.

Manzini, E. (2003). Hong Kong as 'laboratory of the future'. *Design Issues*, *19*(3), 59–7. doi:10.1162/074793603768290847

Margolin, V. (2007). Design, the future, and the human spirit. *Design Issues*, *23*(3), 4–15. doi:10.1162/desi.2007.23.3.4

Margolin, V. (2009). Design in history. *Design Issues*, *25*(2), 94–105. doi:10.1162/desi.2009.25.2.94

Margolin, V. (2010). Doctoral education in Design: Problems and prospects. *Design Issues*, *26*(3), 70–78. doi:10.1162/DESI_a_00031

Marquardt, M. J. (1999). *The global advantage: How world-class organizations improve performance through globalization*. Houston, TX: Gulf Publishing Company.

Martinsons, M. G., & Martinsons, A. B. (1996). Conquering cultural constraints to cultivate Chinese management creativity and innovation. *Journal of Management Development*, *15*(9), 18–35. doi:10.1108/02621719610146239

Mathias, G. (2005). On China design education from 1990 - 2005. *Journal of Package & Design,* (5).

Mayfield, M. (2001). *Thinking for Yourself: Developing Critical Thinking Skills through Readingand Writing* (5th ed.). Thomas Learning.

Mayfield, M. (2011). Innovation. In M. A. Runco & S. R. Pritzker (Eds.), *Encyclopedia of creativity* (2nd ed., pp. 658–666). Academic Press. doi:10.1016/B978-0-12-375038-9.00122-9

McCook, A. (2011). 04 20). Rethinking PhDs. *Nature*, *472*(7343), 280–282. doi:10.1038/472280a PMID:21512549

McCraw, T. K. (2010). *Prophet of Innovation: Joseph Schumpeter and Creative Destruction*. Belknap Press. doi:10.4159/9780674040779

Mckinsey & Company. (2015). *China's iConsumer 2015: A growing appetite for choice and change*. Retrieved from www.iberchina.org/files/Chinas-iConsumer.pdf

McLuhan, M., & Fiore, Q. (2005). *The medium is the massage*. Berkeley, CA: Gingko Press.

Merriam, S. B. (1998). *Qualitative research and case study applications in education. Revised and expanded from "Case study research in education"*. Jossey-Bass Publishers.

Merriam, S. B., Caffarella, R. S., & Baumgartner, L. M. (2007). *Learning in adulthood* (3rd ed.). San Francisco, CA: Jossey- Bass.

Mintzberg, H. (2004). *Managers, not MBA's. A critical vision of company management and management training*. San Francisco, CA: Berret-Koehler.

Misco, T. (2015). *Preparing to Succeed at US Colleges and Universities: A Guidebook for Chinese Students*. Self-published. Printed by CreateSpace Independent Publishing Platform.

Munari, B. (1971). Design as art. London, UK: Penguin Classics.

Munari, B. (1971). *Design as art*. Penguin Books.

Munari, B. (2003). *Bright Minds, Beautiful Ideas: Parallel Thoughts in Different Times*. Consortium Book Sales & Dist.

Murray, R., Caulier-Grice, J., & Mulgan, G. (2010). *The open book of social innovation*. London, UK: The Young Foundation & NESTA.

Nagai, Y., & Noguchi, H. (2003, December). An experimental study on the design thinking process started from difficult keywords: Modeling the thinking process of creative design. *Journal of Engineering Design, VOL.*, *14*(4), 429–437. doi:10.1080/09544820310001606911

Nahavandi, A. (2015). Fostering Creativity and Innovation. SAGE Publications.

Nanfand Daily. (2011). China GDP over Japan became the second largest economy. *Nanfang Daily*. Retrieved April 20, 2016 from http://intl.ce.cn/zgysj/201102/15/t20110215_22215153.shtml

National Advisory Committee on Creative and Cultural Education. (1999). *All Our Futures: Creativity*. Culture and Education.

Newman, J. (2012). *Pebble Smartwatch Pre-Orders Are Sold Out, $10+ Million Pledged*. Retrieved November 12, 2004, from http://techland.time.com/2012/05/10/pebble-smartwatch-pre-orders-sold-out/

Nisbett, R. (2003). *The Geography of Thought: How Asians and Westerners Think Differently … and Why*. New York, NY: Free Press.

Nisbett, R. (2003). *The Geography of Thought: How Asians and Westerners Think Differently…and Why*. New York, NY: The Free Press.

Niu, W. (2006). Development of creativity research in Chinese societies: A comparison of Mainland China, Taiwan, Hong Kong, and Singapore. In J. C. Kaufman & R. J. Sternberg (Eds.), The international handbook of creativity. Cambridge University Press.

Niu, W., & Sternberg, R. J. (2003). Societal and school influences on student creativity: The case of China. *Psychology in the Schools*, *40*(1), 103–114. doi:10.1002/pits.10072

OECD. (1999). *University research in transition*. Organization for Economic Co-operation and Development.

Paletz, S. B. F., & Peng, K. (2008). Implicit theories of creativity across cultures: Novelty and appropriateness in two product domains. *Journal of Cross-Cultural Psychology*, *39*(3), 286–303. doi:10.1177/0022022108315112

Pan, L. S. (2003). Current situation and strategy of development China's higher art education. *Zhuangshi*, (3), 44.

Pang, X. Q. (1987), On arts & crafts: Research of pattern issues. Beijing: Light Industry Publishing House.

Pan, J. (2012). *Report on the China Animation and Comic Education (中国动漫教育发展报告) 2010-2011*. China Social Sciences Press.

Pan, L. S. (2009). Diversity of Design Education in China. *Iridescent-Icograda Journal of Design Research*, *1*(1), 130–136.

Parker, E. (2014). *The Chinese Government Is Investing Heavily in the Maker Movement*. Retrieved May 14, 2014, from http://www.slate.com/blogs/future_tense/2014/05/14/the_chinese_government_is_investing_heavily_in_the_maker_movement.html

Parten, M. (1932). Social participation among preschool children. *Journal of Abnormal and Social Psychology*, *28*(3), 136–147.

Paul, R. (1992, April). Critical thinking: Basic questions and answers. *Think Magazine*.

Paul, C. A. (2005). *Digital Art*. New York, NY: Thames & Hudson.

Paul, R., Elder, L., & Bartell, T. (1997). *California Teacher Preparation for Instruction in Critical Thinking: Research Findings and Policy Recommendations California Commission on Teacher Credentialing, Sacramento California, 1997.* Foundation for Critical Thinking.

Pellegrini, A. D., & Smith, P. K. (2005). *The role of play in human development*. Oxford, UK: Oxford University Press.

Peng, L. (2007). Reflection of contemporary Chinese design education - The status quo of design education and its existing problems in great manufacture country. *Zhuangshi*, (5).

PhD Students, School of Design Hong Kong PolyU. (2016). *PhD Students at The School of Design of the Hong Kong PolyU Facebook Page*. Retrieved from Facebook: https://www.facebook.com/Page-of-the-PhD-Students-at-The-School-of-Design-of-The-Hong-Kong-PolyU-392217560902117/

Piaget, J. (1959). *The language and thought of the child*. Psychology Press.

Picquart, P. (2011). *China in Twenty Years, and the Rest of the World* (E. Salsali, Trans.). Lausanne, Switzerland: Favre.

Pope Edwards, C. (2000). Children's play in cross- cultural perspective: A new look at the six cultures study. *Cultural Research*, *34*(4), 318–338. doi:10.1177/106939710003400402

Qing, Z. (2008). *Status quo analysis and teaching mode exploration of university art design education*. (Master thesis). Hebai Normal University. Retrieved March 10, 2016 from http://cdmd.cnki.com.cn/Article/CDMD-10094-2008125100.htm

Quacquarelli Symonds. (2015). *QS Best Student Cities 2015*. Retrieved November 2, 2015, from http://www.topuniversities.com/city-rankings/2015#sorting=rank+custom=rank+order=desc+search=

Rand, P. (1965). Design and the play instinct. In G. Kepes (Ed.), *Education of Vision* (pp. 156–174). New York, NY: George Braziller Inc.

Richter, W., & Street, W. (2015). China's economic party is 'running out of booze'. Retrieved from www.businessinsider.com.au/chinas-economic-party-is-running-out-of-booze-2015-5

Rittel, H., & Weber, M. D. (1973). Dilemmas in a general theory of planning. *Policy Sciences, 4*(2), 155–169. doi:10.1007/BF01405730

Rivas, M. et al.. (2012). *Tu Cerebro lo es todo (Yourbrainiseverything)*. Plataforma Editorial.

Robinson, K. (2010). *Bring on the Learning Revolution*. Retrieved from http//ted.com/tals/lang/eng/sir_ken_robinson_bring_on_the_revolution.html

Rodber, M., & Wormald, P. (2007). Aligning industrial design education to emerging trends in professional practice and industry. In *DS 43: Proceedings of E&PDE 2007, the 9th International Conference on Engineering and Product Design Education*. University of Northumbria.

Rossi, A. (Director). (2014). *Ivory Tower* [Motion Picture].

Rousseau, J.-J. (1979). *Emile, or on education*. New York, NY: Basic Books.

Rubera, G., & Kirca, A. (2012). Firm innovativeness and its performance outcomes: A meta-analytic review and theoretical integration. *Journal of Marketing, 76*(3), 130–147. doi:10.1509/jm.10.0494

Runco, M. A., & Chand, I. (1995). Cognition and creativity. *Educational Psychology Review, 7*(3), 243–267. doi:10.1007/BF02213373

Russo, J. (2016). Ten reasons China failed: From basic gangsterism to parasitic engagements. Retrieved from Observer website: http://observer.com/2016/02/ten-reasons-china-failed-from-basic-gangsterism-to-parasitic-engagements/

Ryan, J. (2010). 'Chinese learners': misconceptions and realities. In J. Ryan & G. Slethaug (Eds.), International education and the Chinese learner (pp. 37-56). Hong Kong: HKU Press. doi:1.5790/hongkong/978988802845.003.0003

Sakamoto, M., Nakajima, T., Liu, Y., & Todorka, A. (2013). Design and evaluation of micro-crowdfunding: Encourating sustainable behaviour in micro-level crowdfunding. In *AcademicMindTrek '13 Proceedings of International Conference on Making Sense of Converging Media*.

Salas Silva, R. (2008). *Estilos de Aprendizaje a la luz de la Neurociencia (Learning Styles through Neuroscience)*. Bogotá, Colombia: Cooperativa Editorial Magisterio.

Salen, K., & Zimmerman, E. (2003). *Rules of play – game design fundamentals*. Cambridge, MA: MIT Press.

Salili, F. (1996). Accepting personal responsibility for learning. In D. Watkins & J. Biggs (Eds.), *The Chinese learner: cultural, psychological, and contextual influences* (pp. 85–105). Hong Kong: CERC & ACER.

Schiller, F. (2004). *Letters upon the aesthetic education of man, letter XV*. Dover Books.

School of Design, Hong Kong Polytechnic University. (2014). 50th Anniversary Exhibition. Hong Kong: Author.

School of Design, Hong Kong Polytechnic University. (2015). *About PolyU Design*. Retrieved from PolyU Design: http://www.sd.polyu.edu.hk/en/about/about-polyu-design

School of Design, The Hong Kong Polytechnic University. (2015). Hong Kong: School of Design Research Retreat Report.

Schwartz, S. (1999). A theory of cultural values and some implications for work. *Applied Psychology*, *48*(1), 23–47. doi:10.1111/j.1464-0597.1999.tb00047.x

Scott, C. L. (1999). Teachers' biases toward creative children. *Creativity Research Journal*, *12*(4), 321–337. doi:10.1207/s15326934crj1204_10

Sharma, Y. (2011). CHINA: Design education for future economic growth. Retrieved from University World News website: http://www.universityworldnews.com/article.php?story=20111118193754828

Sharp, D. (1987). *Studies in Jungian Psychology by Jungian Analysts*. Inner City Books.

Shierholz, H. (2014). *¿Is there really a shortage of skilled workers?* Economic Policy Institute. Retrieved from http://www.epi.org/publication/shortage-skilled-workers/

Simon, H. (1969). *Sciences of the artificial*. Cambridge, MA: MIT Press.

Sina. (2010). Characteristics of School of Design, Guangzhou Academy of Fine Arts. In *Treat major issues lightly*. Academy Design Graduate Exam Training Institutions. Retrieved April 16, 2016 from http://blog.sina.com.cn/s/blog_61dfedcd0100j543.html

Siu, K. W. M. (2000). Re-construction of learning space for design education. *Design and Education*, *8*(1), 20–28.

Siu, K. W. M. (2003). Nurturing all-round engineering and product designers. *International Journal of Technology and Design Education*, *13*(3), 243–254. doi:10.1023/A:1026101210221

Siu, K. W. M. (2004). Flexible learning environment for design education: Meeting the changes. In D. Saunders, K. Brosnan, M. Walker, A. Lines, J. Storan, & T. Acland (Eds.), *Learning transformations: Changing learners, changing organisations, changing communities* (pp. 110–113). London: Forum for the Advancement of Continuing Education, University of East London.

Siu, K. W. M. (2009). New trends of research in postgraduate design education in China. *US-China Education Review*, *6*(9), 16–21.

Siu, K. W. M. (2010). Meeting the new needs: Design research education in China. *Research in Higher Education Journal*, *6*(March), 157–173.

Siu, K. W., & Contreras, G. (2015). Personal Interaction and Informal Learning: The Case of China. In P. C. O. Mejiuni & P. C. O. Mejiuni (Eds.), *Measuring and Analyzing Informal Learning in the Digital Age* (pp. 140–152). Hershey, PA: IGI Global. doi:10.4018/978-1-4666-8265-8.ch010

Smolucha, L., & Smolucha, F. (1986). L. S. Vygotsky's theory of creative imagination. In *Proceedings of the Annual Convention of the American Psychological Association.*

Sorabjee, D. (2012). China has 400 plus design schools, much less exist in India. *The Economic Times*. Retrieved from http://economictimes.indiatimes.com/industry/china-has-400-plus-design-schools-much-less-exist-in-india/articleshow/12241365.cms

Staats, L. (2011). The Cultivation of Creativity in the Chinese Culture—Past, Present, and Future. *Journal of Strategic Leadership.*, *3*(1), 45–53.

Staats, L. K. (2011). The cultivation of creativity in the Chinese culture—Past, present, and future. *Journal of Strategic Leadership*, *3*(1), 45–53.

Standard Ranking Research Institute. (2015). *Undergraduate Salary Ranking*. Retrieved from http://edu.163.com/15/0428/14/AO9UO15E00294MP6.html

Stein, M. I. (1953). Creativity and culture. *The Journal of Psychology*, *36*(2), 311–322. doi:10.1080/00223980.1953.9712897

Sternberg, R. J., & Lubart, T. I. (1995). *Defying the crowd: Cultivating creativity in a culture of conformity*. Free Press.

Sternberg, R. J., & Williams, W. M. (1996). *How to Develop Student Creativity*. ASCD.

Su, M. Z., & Huang, K. (2012). Practice ability reform and training for art and design. *Examination (Teaching & Research)*. Retrieved April 12, 2016 from http://www.xzbu.com/9/view-3413912.htm

Sun, F., & Shen, Y. (2013). The influence of international modernism on China's hundred-year design. *Idea & Design*, (4). Retrieved March 15, 2016 from http://www.cnki.com.cn/Article/CJFDTotal-CYJY201304007.htm

Sun, L. (孙立军) (Ed.) (2005). The Situation of Local Animation Education (国内动画教育状况). China Animation Industry Annual Report 2005 (pp. 112-137). China: Ocean Press.

Sutton-Smith, B. (1997). *The ambiguity of play*. Cambridge, MA: Harvard University Press.

Svoboda, S. (2015, June 2). *Why do so many Chinese students choose US universities?* Retrieved from BBC News: http://www.bbc.com/news/business-32969291

Tang, C. M. (2010). *Three values of Confucian culture on contemporary environmental art design*. (Master thesis). Northeast Normal University. Retrieved March 10, 2016 from http://cdmd.cnki.com.cn/Article/CDMD-10200-2010179169.htm

Tang, M. X. (2016). Towards an Outcome-Based Postgraduate Education: A perspective for the students. Hong Kong: Academic Press.

Tao, L. (陶琳娜) (2010). The Requirements of Teachers in Film and Television Education (影视动画教育对教师的要求). *Master*. 163.

Tatlow, D. K. (2014, September 14). *Q. and A.: Yong Zhao on Education and Authoritarianism in China*. Retrieved from The New York Times blog: http://sinosphere.blogs.nytimes.com/2014/09/14/q-and-a-yong-zhao-on-education-and-authoritarianism-in-china/?_php=true&_type=blogs&_php=true&_type=blogs&_r=2&

The Economist. (2013). *Crash Course*. Retrieved on 24 November, 2015 from: http://www.economist.com/news/schoolsbrief/21584534-effects-financial-crisis-are-still-being-felt-five-years-article

The Free Dictionary. (2016). Retrieved from http://www.thefreedictionary.com/design

The New York Times. (1989). *What's in a name?* Retrieved from http://www.nytimes.com/1989/04/02/magazine/what-s-in-a-name.html?pagewanted=all

The State Council. (2015). 'Made in China 2025' plan issued Retrieved from The State Council The People's Republic of China website: http://english.gov.cn/policies/latest_releases/2015/05/19/content_281475110703534.htm

Tischler, L. (2011). China knows it must innovate. But can it? Retrieved from Fastco Design website: http://www.fastcodesign.com/1664945/china-knows-it-must-innovate-but-can-it

Tong, H. M. (2009). Introduction: re-cognition of foundational course. In Zhao & Feng (Eds.), Concept: Understanding and expression. Guangzhou: Linnan Fine Arts Publishing House.

Tong, H. M. (2010). Growth of Chinese design. *21 Century Business Review*, (2). Retrieved March 12, 2016 from http://www.cnki.com.cn/Article/CJFDTOTAL-ESYS201002044.htm

Tongji University. (2013). College of Design and Innovation. In *College manifesto*. Tongji University. Retrieved April 18, 2016 from http://tjdi.tongji.edu.cn/about.do?ID=55&lang=_en

Tse, T., & Esposito, M. (2014). *Academia is disconnected from the real world.* Retrieved from Financial Times: http://www.ft.com/cms/s/2/4f5fc7a2-7861-11e3-831c-00144feabdc0.html#axzz3uGO9Zo2H

Turner, M. (2016). Something terrible seems to be happening to the Chinese job market. Retrieved from Business Insider website: http://www.businessinsider.com/china-job-postings-declining-since-november-2016-

Varun, A. (2014). Article. *International Journal of Education and Psychological Research*, *3*(3), 49.

Vázquez, J. (2016). *Cincotópicossobreinnovacióndocente (Five topics about teaching innovation).* Retrieved from http://www.universidadsi.es/cinco-topicos-innovacion-docente/

Vigleik, N. (n.d.). *An introduction to form-giving in Chinese product design*. Academic Press.

Vision Union. (2015). Origin: Design Education and Innovation in the 'New Normal' Society - 2015 National Design Education Academic Forum & 'Design Subcommittee, Chinese Association of Higher Education' Annual Conference. *Vision Union*. Retrieved April 10, 2016 From http://www.visionunion.com/article.jsp?code=201511030002

Volet, S. E., & Renshaw, P. D. (1996). Chinese students at an Australian university: adaptability and continuity. In D. A. Watkins & J. B. Biggs (Eds.), *The Chinese learner: Cultural, psychological and contextual influences* (pp. 205–220). Hong Kong: HKU Press.

Von Oech, R. (1992). *Creative whack pack*. Stamford, CT: US Games Systems.

Wang, C. W. (1963). Love to see and hear. Authors Publishing House.

Wang, K. (王可) (2012). Thinking of a few questions in the Higher Animation Education (关于中国高校动漫教育若干问题的思考). *Art in China*, (2), 115-116.

Wang, M. Z. (2000). Integration, combination and foreseen: on disciplines alteration of Academy of Arts & Design, Tsinghua University. *Arts Observation*, (6), 5.

Wang, S. Z. (2008a). Critical analysis of China design education. *Sina Blog*. Retrieved March 8, 2016 from http://blog.sina.com.cn/s/blog_4bdabb4901007t5o.html

Wang, S. Z. (2008b). Critical analysis of China design education. *Sina Blog*. Retrieved March 8, 2016 from http://blog.sina.com.cn/s/blog_4bdabb4901007t5p.html

Wang, Y. X. (1998). Principle of the development strategies of higher education. *Higher Education Research,* (1), 20.

Wang, Y. X., & Yue, X. Y. (2011). Upgrading of arts and its significance - Professor Wang Tingxin interviews. *The Art Community,* (2). Retrieved April 15, 2016 from http://www.cnki.com.cn/Article/CJFDTotal-YSXJ201102003.htm

Wang, Z. (王正友) (2007). An In-depth Study of the Current Situation of China Animation Education (中国动画教育现状探析). *Education and Vocation*, (9), 103-104.

Wang, J. (2004). The global reach of a new discourse: How fast can 'Creative Industries' travel? *International Journal of Cultural Studies*, *7*(1), 9–19. doi:10.1177/1367877904040601

Wang, S. Z. (1995). *Modern design history of the world*. Shenzhen: New Century Publishing House.

Wang, T., Zhang, Y., Pen, J., Lu, Q. Z., & Jin, K. (Eds.). (2011). *Unlimited territory: 2011 experimental teaching and learning for graduate design projects in environmental art by four schools' supervisors*. Beijing: China Architectural Industry Publishing House.

Wang, X. J. (2009). Contemporary Forces Driving the Development of Design. *Iridescent-Icograda Journal of Design Research*, *1*(1), 138–146.

Watkins, D. A., & Biggs, J. B. (1996). *The Chinese learner: cultural, psychological and contextual influences*. Hong Kong: HKU Press.

Watts, A. (2014). *Gruber Thinking in Climate Science: Disconnect Between Academia And The Real World*. Retrieved from wattsupwiththat.com: http://wattsupwiththat.com/2014/12/12/gruber-thinking-in-climate-science-disconnect-between-academia-and-the-real-world/

Weisul, K. (2014). *In Crowdfunding, All-or-Nothing Campaigns Are More Successful*. Retrieved November 12, 2014, from http://www.inc.com/kimberly-weisul/why-your-crowdfunding-campaign-should-be-all-or-nothing.html?cid=em01011week25day19d

Wei, X. H., & Li, Q. Y. (2013). The Confucian Value of Harmony and its Influence on Chinese Social Interaction. *Cross-Cultural Communication*, *9*(1), 60–66.

Wertime, D. (2015). *Chinese netizens to Fiorina: You're right, We don't innovate: Many in China have been complaining about their education system for years*. Retrieved on 7th May, 2016: http://foreignpolicy.com/2015/06/01/chinese-netizens-agree-fiorina-innovation-comment-education-system/

Westby, E. L., & Dawson, V. L. (1995). Creativity: Asset or burden in the classroom? *Creativity Research Journal*, *8*(1), 1–10. doi:10.1207/s15326934crj0801_1

Wiens, K. (2016). Huawei just copied the iPhone—Down to the Last Screw. Retrieved from Wired website: http://www.wired.com/2016/05/huawei-iphone-screws-ifixit/

Wilber, K. (2010). Introduction to integral theory and practice: IOS basic and the AQAL map. *AQAL Glossary, 1*(3). Retrieved December 12, 2015 from http://www.humanemergence.nl/uploads/2011/03/IOS-Basic-Intro-to-Integral.pdf

Wilber, K. (1995). *Sex, ecology, spirituality: the spirit of evolution*. Boston, MA: Shambhala.

Williams, A. (2015). *Swinburne and Hong Kong Poly University partner to offer joint PhD program*. Retrieved from Swinburne University of Technology: http://www.swinburne.edu.au/news/latest-news/2015/11/swinburne-and-hong-kong-poly-university-partner-to-offer-joint-phd-program.php

Willis, J. (2012). *A Neurologist Makes the Case for Teaching Teachers About the Brain*. Retrieved from http://www.edutopia.org/blog/neuroscience-higher-ed-judy-willis

Wilson, M. (2015). Why China is investing In design. Retrieved from Fastco Design website: http://www.fastcodesign.com/3041089/why-china-is-investing-in-design

Wong, S.-L. (2015). Hangzhou shows the way to China's new economy. Retrieved from Reuters website: http://www.reuters.com/article/us-china-economy-hangzhou-idUSKBN0O52NO20150521

Wong, W. (1973). *Principles of two-dimensional design*. New York, NY: Van Nostrand Reinhold.

Wong, W. (1977). *Principles of three-dimensional design*. New York, NY: Van Nostrand Reinhold.

Wong, Y. L., & Siu, K. W. M. (2012). Is there creativity in design? From a perspective of school design and technology in Hong Kong. *Asia Pacific Education Review*, *13*(3), 465–474. doi:10.1007/s12564-012-9208-y

Wu, Q. (2006). Creative industries and innovation in China. *International Journal of Cultural Studies*, *9*(3), 263–266. doi:10.1177/1367877906066872

Wu, W. P. (2010). *Higher Education Innovation in China*. Richmond, VA: Virginia Commonwealth University.

Wu, X. F. (2007). *Want to do design: Training program to rapidly enhance innovative capacity*. China Social Science Academic Press.

Xiao, Y. (2011). An Analysis of the Current Situation in China Animation Education. *Journal of Beijing Union University*, *9*(3), 71–77.

Xin, F. (2011). *Industrial design: Contrasting the United States and Chinese methods from the perspective of an industrial designer who has both studied and worked in the U.S. and China. (Master of Design)*. University of Cincinnati.

Xin, H. Q. (1999). *The Method of shape & form formulation*. Hangzhou: China Academy of Fine Arts Publishing House.

Xu, C., & Cui, B. S. (2010). *A Discourse on Current Situation and Strategy of University Animation Education*. Heilongjian Science and Technology Information.

Xueqin, J. (2011, July 19). Thinking Right: Coaching a Wave of Chinese Students for College in America. *The Chronicle of Higher Education*. Retrieved March 16, 2016, from http://chronicle.com/article/Thinking-Right-Coaching-a/128287/

Yang, C. Y., & Lin, W. L. (1992). Lei Guiyuan on pattern art. Hangzhou: Zhejiang Academy of Fine Arts Publishing House.

Yang, I. K. Monique. (2016). Mapping China: Design - Education. Retrieved from Dutch Culture website: http://dutchculture.nl/en/mapping/mapping-china-design-education

Yang, W. Y. (1993). Confucius. UNESCO.

Yang, X. (2016). Yang Xiaoyang. *People Weekly,* (10). Retrieved March 16, 2016 from http://www.wtoutiao.com/p/10btIOK.html

Ye, J. (2010). The Current Situation in China Animation Education, an Analysis of the Problems and Development Strategies. *Journal of Hubei University of Economics*, *7*(8), 174–175.

Yeo, J. P.-H. (2012). Research in Art and Design Education: A review. *Academic Research International*, *3*(3), 303.

Yin, R. K. (2013). *Case study research: Design and methods*. Sage Publications.

Yu, Q. (2006). General theory of design science. Chongqing: Chongqing University Publishing House.

Yuan, X. Y. (2004). Research of development path of China art design. Beijing: Beijing University of Technology Publishing House.

Žegarac, V. (2007). A cognitive pragmatic perspective on communication and culture. In H. Kotthoff & H. Spencer-Oatey (Eds.), *Handbook of Intercultural Communication* (pp. 31–53). Berlin: Walter de Gruyter.

Zhang, L., & Dou, J. H. (2011). *Reanalysis of Industrial Design Education in China*. Retrieved from http://dolcn.com/data/cns_1/article_31/paper_311/pind_3111/2011-06/1308229464.html

Zhang, L., & Sternberg, R. (2012). *Culture and Intellectual Styles*. New York, NY: Springer.

Zhao, Y. (2012). lunking innovation and creativity. *Kappan, 49*(1).

Zhen, L. (2006). *International Classroom Culture.* Centre for teaching & Learning resources. Retrieved from http: //www.chinabuddhismencyclopedia.com/en/index.php/Kalama_Sutta

Zheng, S. Y., Nie, Y., Zhang, B. M., & Tang, L. T. (2010). The Development strategy research of China' art design education. Beijing: Tsinghua University Publishing House.

Zheng, X. N. (2009). Establishing innovative education mode for art & design universities. *China Higher Education*, (21), 56.

Zheng, Y., & Gao, X. (2015). Research and practice of innovative talent training mechanism of Comprehensive art colleges and universities. *China University Teaching*, (12), 22.

Zhou, X. (2013). Philosophical reflections on contemporary design concepts. *Zhuangshi*, (6).

Zhuge, K. (2007). Inheritance from the fission. Chongqing: Chongqing University Publishing House.

Zimmerman, J. (2006). *Innocents abroad: American teachers in the American century*. Cambridge, MA: Harvard University Press. doi:10.4159/9780674045453

Zuo, M. (2015, May 29). 8000 Chinese students in US expelled for low grades, academic dishonesty, breaking rules. *South China Morning Post*. Retrieved March 22, 2016, from http://www.scmp.com/news/china/money-wealth/article/1812115/roughly-3pc-chinese-students-us-expelled-low-grades-academic

About the Contributors

Kin Wai Michael Siu is Chair Professor of Public Design. He is Founder and Lab Leader of the Public Design Lab, School of Design, The Hong Kong Polytechnic University. He was the Visiting Scholar of the University of Cambridge, Fulbright Scholar of the MIT, ASIA Fellow of the National University of Singapore, and Visiting Scholar of the UC Berkeley. He has served the Hong Kong Government's advisory committees and professional bodies as members. He owns more than 50 invention and design patents. His research interests are in design culture, public design, universal and inclusive design, and user-reception.

Giovanni J. Contreras Garcia is currently working towards his PhD in Design Education at the School of Design of The Hong Kong Polytechnic University. He has been design lecturer and practicing industrial designer in China, the US and Mexico. He was Fulbright Scholar at the University of Cincinnati where he obtained a master of design, and a recipient of the Hong Kong PhD fellowship. He has been member of the Industrial Designers Society of America and winner of the China iF Design Award. His research interests sit at the intersection of philosophy, design, education, and digital technologies.

***

Stefano Ceppi graduated in Product Design from Central Saint Martin's University, in London. As final project he developed a tilting car in collaboration with the Royal College of Art (London, 2004). He works as product designer and consultant for various vehicles production companies. He has worked as a lecturer in Shanghai (RDI Donghua University Campus) teaching Critical Thinking and many other subjects in PD; he was also responsible for the Industrial Design modules for Quasar Design College in Rome, Italy. His major areas of interests are advanced design and innovation technologies across various ranges of skill. He participated as advanced designer in the design of the first e-bike in Piaggio's history.

Yu-Chen Huang received her Ph.D. in Graduate School of Design, National Yunlin University of Science and Technology, Taiwan in 2014. In 2015 she was hired as the Assistant Professor at Department of Creative Design and Architecture, National Kaohsiung University. Her publication and professional field are Sustainable Design, Design Development, Design Strategy, Cultural and Creative Design.

Jiabei Jiang received her BA and MA in industrial design from Politecnico di Milano respectively in 2009 and 2012. After that she was hired as a lecturer in industrial design by the School of Industrial Design at Nanjing University of the Arts. Her research interests includes product design, design engineering, service design and interaction design.

Rémi Leclerc graduated with a Masters in Industrial Design from Les Ateliers - ENSCI in Paris. Having lived in Asia, Africa, America, and Europe, his understanding of the representations of these cultures invigorate and nurture his design research perspectives. His work is anchored in his exploration of different notions of play and conventions of interactivity. He draws on the characteristics of traditional and contemporary objects to develop prospective play products and systems. He has worked collaboratively on projects with socially responsible organizations in Hong Kong, Paris and New York. Rémi leads HK PolyU Design's Undergraduate Scheme and its PolyPlay research unit, and is a member of the International Toy Research Association. PolyPlay reserach work investigates the relationships between design, play, and education, and bridges scholarship, community, and industry, to bring to life innovative products and systems.

Henry Ma is an Associate Dean and Associate Professor of School of Design of The Hong Kong Polytechnic University and Guest Research Professor of the Beijing Film Academy. He received the "Creative Industry Leader Award in China" and the Hong Kong Digital Entertainment Industry Person of the Year Award (Digital Entertainment Education) in 2012 and 2007 respectively. Henry graduated from the Hong Kong Academy for Performing Arts School of Technical Arts, School of Management and Centre for Mass Communication Research at University of Leicester, and received his PhD from School of Design, The Hong Kong Polytechnic University. Henry participated in a number of research projects as principle investigators, project leaders or deputy project leaders with a cumulative sum of funding over fifteen million Hong Kong dollars. These research projects include study of teaching for creativity, management of creativity, project-based blended learning strategy, game-based learning, augmented reality and motion capture.

Miguel Rivas Spanish born in Argentina, Prof. Rivas has been studying or working in Buenos Aires, Lausanne, London, Luanda, Madrid, Rio de Janeiro, Shanghai and Washington, DC. He holds a BSc (Business and Economics) and postgraduate studies in Strategic Marketing and Applied Neuroscience to institutions. Currently, Prof. Rivas is the CEO of BricSense Ltd. and director of BricSense Academy, Director/Representative at the European Union with Alta Gerencia Internacional and a Certified Quality & Performance Senior Consultant at UNWTO. He is also a visiting professor with University of Central Lancashire (UCLan, UK), in charge of Change and Development of Asia-Pacific, Madrid Campus (Kingdom of Spain). He is a co-author of "Tu Cerebro lo es todo" (Your brain is everything), 2011, endorsed by Antonio Damasio, Ph D and published by Plataforma Editorial, Barcelona.

Rebecca Ruige Xu currently teaches computer art and animation as an Associate Professor at Syracuse University. Her artwork and research interests include artistic data visualization, interactive design, experimental animation, visual music, digital performance and virtual reality. Her recent work has been appeared at: Ars Electronica Animation Festival; SIGGRAPH Art Gallery; Aesthetica Short Film Festival, UK; Museum of Contemporary Art, Italy; Los Angeles Center for Digital Art; Magmart International Videoart Festival, Italy; FILE - Electronic Language International Festival, Brazil; Techfest -Technical Arts Exhibition, India; Colloquium culture and digitization, Switzerland; CYNETart, Germany; International Digital Art Exhibition, China; Huddersfield Festival, UK; Boston Cyberarts Festival.

Edmond Salsali is an instructor, digital artist, and researcher based in New York City. He has been a professional multimedia designer and developer for many years and has taught at different colleges and universities in the US, with courses in multimedia, animation, 2D and 3D interactivity, game design, Web design, and graphic design. His work has been featured in individual and group exhibitions, competitions, and festivals worldwide. He is currently an Assistant Professor of Digital Arts and Chair of the Communication, Graphic Design & Multimedia Department at Georgian Court University in Lakewood, NJ. Edmond holds a Ph.D. in Digital Arts from the University of Paris VIII, where he studied with Professors Edmond Couchot, Christine Buci-Glucksmann, Michel Bret, Hervé Huitric, and other eminent pioneers in the digital and interactive art fields. His current research and practice incorporate the use of virtual reality systems and their artistic implements, 3D game design and development, and dynamic interactivity for online and offline 3D virtual environments.

Jui-Che Tu received his Ph.D. in Industrial Education, University of Tennessee, USA in 1991. In 2004 he was hired as the Professor at Department of Industry Design, Da-Yeh University and is currently the Head of the Graduate School of Design, National Yunlin University of Science and Technology in Taiwan. His publication and professional field are Industrial Design, Green Design, Advanced Life-Style Design, Design Education, Cultural and Creative Design.

Fang Xu is an Associate Professor in Art & Design, UNSW in Australia, where he is the Coordinator of Environments / Spatial Design, the supervisor of PhD and Master of Design students. Fang's expertise is drawn from his cross-cultural background, wide range of professional design experience and interdisciplinary research interests. With more than thirty years experience as a practising designer and tertiary educator in Australia and China, Fang has a deep understanding of the nature of design and its social influences and cultural values, which shape contemporary design education and practice. As an educator, Fang's interest is in the relationship between culture, design process and design methodology, and the influence of the manifestation of culture on design teaching and learning in the context of Australian and Asian Universities. As a researcher, his interest is in design's increasingly hybrid and inter-disciplinary approach as it responds to the transformation of the design industry in the real world.

Li Zeng is an Assistant Professor of Graphic Design at University of Central Arkansas, and founder of Plus Designer. Her research focuses on design thinking, service design and human centered design methodology. She has been recognized by some major competitions and publications including: Adobe Award, How Top 10 Poster Design Award, Design China, Visual China, Vision-Union and much more. She received a B.A from China Central Academy of Fine Art and M.F.A from Minneapolis College of Art and Design.

Ding Zhou earned his BA in industrial design from Zhejiang Sci-Tech University in 2006. After his graduation with the MA in art design from Shandong University in 2009, he was hired as a lecturer in industrial design by the School of Industrial Design at Nanjing University of the Arts. In 2012, he was sponsored by Jiangsu province government to work as a bilingual visiting staff at the University of Virginia, US. Further, as a visiting scholar in industrial design with the Chinese national scholarship, he worked at the School of Design of Victoria University of Wellington from 2014 to 2015 in New Zealand. He currently dedicates to innovative design applications of additive manufacturing. Design education and tangible interaction design also are included in his research interests.

Index

O

P

Q

R

S

T

V

W

www.ingramcontent.com/pod-product-compliance
Ingram Content Group UK Ltd.
Pitfield, Milton Keynes, MK11 3LW, UK
UKHW050946290926
15470UKWH00005B/1668